T0214970

Lecture Notes in Computer Science   10940

Commenced Publication in 1973
Founding and Former Series Editors:
Gerhard Goos, Juris Hartmanis, and Jan van Leeuwen

Editorial Board

David Hutchison
  Lancaster University, Lancaster, UK
Takeo Kanade
  Carnegie Mellon University, Pittsburgh, PA, USA
Josef Kittler
  University of Surrey, Guildford, UK
Jon M. Kleinberg
  Cornell University, Ithaca, NY, USA
Friedemann Mattern
  ETH Zurich, Zurich, Switzerland
John C. Mitchell
  Stanford University, Stanford, CA, USA
Moni Naor
  Weizmann Institute of Science, Rehovot, Israel
C. Pandu Rangan
  Indian Institute of Technology Madras, Chennai, India
Bernhard Steffen
  TU Dortmund University, Dortmund, Germany
Demetri Terzopoulos
  University of California, Los Angeles, CA, USA
Doug Tygar
  University of California, Berkeley, CA, USA
Gerhard Weikum
  Max Planck Institute for Informatics, Saarbrücken, Germany

More information about this series at http://www.springer.com/series/8637

Abdelkader Hameurlain · Roland Wagner (Eds.)

Transactions on Large-Scale Data- and Knowledge-Centered Systems XXXVII

 Springer

Editors-in-Chief
Abdelkader Hameurlain
IRIT, Paul Sabatier University
Toulouse
France

Roland Wagner
FAW, University of Linz
Linz
Austria

ISSN 0302-9743   ISSN 1611-3349   (electronic)
Lecture Notes in Computer Science
ISSN 1869-1994   ISSN 2510-4942   (electronic)
Transactions on Large-Scale Data- and Knowledge-Centered Systems
ISBN 978-3-662-57931-2   ISBN 978-3-662-57932-9   (eBook)
https://doi.org/10.1007/978-3-662-57932-9

Library of Congress Control Number: 2018950529

© Springer-Verlag GmbH Germany, part of Springer Nature 2018
This work is subject to copyright. All rights are reserved by the Publisher, whether the whole or part of the material is concerned, specifically the rights of translation, reprinting, reuse of illustrations, recitation, broadcasting, reproduction on microfilms or in any other physical way, and transmission or information storage and retrieval, electronic adaptation, computer software, or by similar or dissimilar methodology now known or hereafter developed.
The use of general descriptive names, registered names, trademarks, service marks, etc. in this publication does not imply, even in the absence of a specific statement, that such names are exempt from the relevant protective laws and regulations and therefore free for general use.
The publisher, the authors and the editors are safe to assume that the advice and information in this book are believed to be true and accurate at the date of publication. Neither the publisher nor the authors or the editors give a warranty, express or implied, with respect to the material contained herein or for any errors or omissions that may have been made. The publisher remains neutral with regard to jurisdictional claims in published maps and institutional affiliations.

This Springer imprint is published by the registered company Springer-Verlag GmbH, DE
part of Springer Nature
The registered company address is: Heidelberger Platz 3, 14197 Berlin, Germany

Preface

This volume contains five fully revised regular papers, covering a wide range of very hot topics in the field of data security in clouds, privacy languages, probabilistic modelling in linked data integration, business intelligence based on multi-agent systems, collaborative filtering, and prediction accuracy.

We would like to sincerely thank the editorial board for thoroughly reviewing the submitted papers and ensuring the high quality of this volume.

Special thanks go to Gabriela Wagner for her availability and her valuable work in the realization of this volume of the Transactions on Large-Scale Data- and Knowledge-Centered Systems.

June 2018

Abdelkader Hameurlain
Roland Wagner

Organization

Editorial Board

Reza Akbarinia	Inria, France
Bernd Amann	LIP6 – UPMC, France
Dagmar Auer	FAW, Austria
Djamal Benslimane	Lyon 1 University, France
Stéphane Bressan	National University of Singapore, Singapore
Mirel Cosulschi	University of Craiova, Romania
Tran Khanh Dang	Ho Chi Minh City University of Technology, Vietnam
Dirk Draheim	Tallin University of Technology, Estonia
Johann Eder	Alpen Adria University Klagenfurt, Austria
Anastasios Gounaris	Aristotle University of Thessaloniki, Greece
Theo Härder	Technical University of Kaiserslautern, Germany
Sergio Ilarri	University of Zaragoza, Spain
Petar Jovanovic	Universitat Politècnica de Catalunya, BarcelonaTech, Spain
Dieter Kranzlmüller	Ludwig-Maximilians-Universität München, Germany
Philippe Lamarre	INSA Lyon, France
Lenka Lhotská	Technical University of Prague, Czech Republic
Liu Lian	University of Kentucky, USA
Vladimir Marik	Technical University of Prague, Czech Republic
Jorge Martinez Gil	Software Competence Center Hagenberg, Austria
Franck Morvan	Paul Sabatier University, IRIT, France
Kjetil Nørvåg	Norwegian University of Science and Technology, Norway
Themis Palpanas	Paris Descartes University, France
Torben Bach Pedersen	Aalborg University, Denmark
Günther Pernul	University of Regensburg, Germany
Soror Sahri	LIPADE, Paris Descartes University, France
A Min Tjoa	Vienna University of Technology, Austria
Shaoyi Yin	Paul Sabatier University, Toulouse, France

Contents

Keeping Secrets by Separation of Duties While Minimizing the Amount
of Cloud Servers.   1
    Ferdinand Bollwein and Lena Wiese

LPL, Towards a GDPR-Compliant Privacy Language:
Formal Definition and Usage 41
    Armin Gerl, Nadia Bennani, Harald Kosch, and Lionel Brunie

Quantifying and Propagating Uncertainty in Automated Linked
Data Integration 81
    Klitos Christodoulou, Fernando Rene Sanchez Serrano,
    Alvaro A. A. Fernandes, and Norman W. Paton

A Comprehensive Approach for Designing Business-Intelligence
Solutions with Multi-agent Systems in Distributed Environments 113
    Karima Qayumi and Alex Norta

Enhancing Rating Prediction Quality Through Improving the Accuracy
of Detection of Shifts in Rating Practices. 151
    Dionisis Margaris and Costas Vassilakis

Author Index 193

Keeping Secrets by Separation of Duties While Minimizing the Amount of Cloud Servers

Ferdinand Bollwein[1] and Lena Wiese[2](✉)

[1] Institute of Computer Science, TU Clausthal, Clausthal-Zellerfeld, Germany
ferdinand.bollwein@tu-clausthal.de
[2] Institute of Computer Science, University of Goettingen, Göttingen, Germany
wiese@cs.uni-goettingen.de

Abstract. In this paper we address the problem of data confidentiality when outsourcing data to cloud service providers. In our separation of duties approach, the original data set is fragmented into insensitive subsets such that each subset can be managed by an independent cloud provider. Security policies are expressed as sets of confidentiality constraints that induce the fragmentation process. We assume that the different cloud providers do not communicate with each other so that only the actual data owner is able to link the subsets and reconstruct the original data set. While confidentiality is a hard constraint that has to be satisfied in our approach, we consider two further optimization goals (the minimization of the amount of cloud providers and the maximization of utility as defined by visibility constraints) as well as data dependencies that might lead to unwanted disclosure of data. We extend prior work by formally defining the confidentiality and optimization requirements as an optimization problem. We provide an integer linear program (ILP) formulation and analyze different settings of the problem. We present a prototype that exploits a distributed installation of several PostgreSQL database systems; we give an in-depth account of the sophisticated distributed query management that is enforced by defining views for the outsourced data sets and rewriting queries according to the fragments.

1   Introduction

Data outsourcing and data processing in cloud services now-a-days manifest in different variants and for different use cases: numerous providers offer cloud services where data is processed off-premise and is no longer under the control of the actual data owner. Using such cloud services offers several advantages like:

- scalability: customers can book resources according to their demand leading to a reduction of hardware and maintenance costs;
- flexibility: cloud servers can run with an optimal payload and hence the overall consumption of energy can be reduced;
- availability: customers can access their data from anywhere – independent of their physical location;

© Springer-Verlag GmbH Germany, part of Springer Nature 2018
A. Hameurlain and R. Wagner (Eds.): TLDKS XXXVII, LNCS 10940, pp. 1–40, 2018.
https://doi.org/10.1007/978-3-662-57932-9_1

– reliability: cloud providers offer service level agreements and invest in reliability of their systems so that data loss is reduced.

In particular for large data sets or when data can be accessed by many different parties, using cloud services for data outsourcing offers many benefits. For the purposes of this article we categorize data outsourcing into three different variants:

– **Data storage** (single-owner read and write access): a single data owner manages his or her data remotely at outsourcing providers; the data owner has read and write access on the data. As one example for this setting, businesses can profit by sparing the cost of maintaining an own computing center; as another example, a patient can maintain his or her personal electronic health record in the cloud.
– **Data publishing** (single-owner write, multiple-user read access): a single data owner stores some data at outsourcing providers. Other users can then query the data (that is, with read-only access). Access control may be enforced on the published data: certain data items may only be queried by privileged users. A particular variant of data publishing is the statistical evaluation of the outsourced data: in this case, the data can even be distorted as long as the final evaluation on the distorted data does not diverge much from the evaluation on the original data.
– **Data sharing** (multiple-owner read and write access): multiple data owners manage their data collaboratively where each data owner may selectively be entitled only to certain read and write accesses. In the business example, in contrast to data storage, different business units may share data and have the added value of executing fine-grained access control; in the electronic health record example, patients can selectively allow read and write access to some of their health data for medical personnel.

However, outsourcing data also means that the user loses control over the data and the cloud providers have to be trusted in order to ensure confidentiality of sensitive or business-critical data. The three outsourcing variants each have different security requirements which we now briefly discuss.

When using the *data storage* variant, data should be inaccessible to other users or the cloud service provider. One possibility to ensure confidentiality would be to encrypt the data stored in the cloud database, however, this limits the provider's ability of processing the data to answer complex queries by the user. To achieve query answering on ciphertexts, property-preserving encryption schemes are available that allow to sort ciphertexts, or to search for encrypted keywords on the ciphertexts. While these schemes enable certain operations on the encrypted data they come at the cost of an increased overhead; prototype systems that apply such encryption schemes are [27,31] for SQL databases (which also feature homomorphic encryption for aggregations) and [34] for NoSQL databases.

When using the *data publishing* variant, only certain characteristics of the data are confidential. These characteristics can be hidden by distorting the data with some noise. Examples for the publishing of distorted data include k-anonymity [28] and differential privacy [20]. For the example use cases, business-critical or medical data would only published in a curated version including slightly modified or generalized values.

In case of the *data sharing* variant, data should selectively be accessible to some users but should be hidden from other users or third parties. For the purpose of encrypted data sharing with multiple owners, *multi-user* property-preserving encryption schemes exist but the distribution of appropriate cryptographic keys is a major complication.

As an alternative, this article proposes a *separation of duties* approach to address the complexity of encryption – however, we want to reinforce that in a real-world system a combination of separation of duties and encryption would be of added practical value. Our proposed approach applies to the data outsourcing variant of data storage, where data confidentiality is achieved by distributing data fragments among different separate cloud service providers – with a specific focus on data storage in cloud databases ("database-as-a-service"). If in addition selective access rights on the distributed fragments are given to several users, the data outsourcing variant of data sharing can also profit from our separation of duties approach. We refrain however from distorting data as would be necessary for some data publishing scenarios mentioned above. A particular scenario that our approach is suited for is the outsourcing of medical data into repositories. Here several data providers can submit their data into these repositories. The important difference to encryption-based approaches is that statistical evaluations are still possible on plaintext data; support for statistics on certain attribute combinations can be enforced by visibility constraints.

Generally, the term separation of duties means that a specific task is handled by multiple entities to prevent malicious behavior which could be carried out by a single entity in control of the whole task. In the context of preserving confidentiality in cloud databases, this is based on the observation that in many scenarios data only becomes sensitive in association with other data. By distributing the data among multiple database servers with a technique called *vertical fragmentation*, these sensitive associations can be broken up such that each server only maintains an insensitive portion of the data. The fragmentation process is guided by a security policies that contains so-called confidentiality constraints. As long as the servers are not collaborating to reestablish the association, this separation of duties approach ensures the confidentiality of the underlying data. The huge advantage of our approach is that the data can be outsourced in plaintext because and the data do not have to be encrypted which would drastically limit the servers' ability to process client queries. Moreover, no noise is added and the data outsourced plaintext data retain their original values.

It is a challenging task to decide how the data is distributed among the servers. On the one hand, the security requirements have to be met but on the other hand the number of involved servers should be relatively small to limit the costs for the user and ensure efficient querying. Therefore, the proposed separation of duties approach can be viewed as a typical mathematical optimization problem. In this article we consolidate and extend our prior work [6,7] in a multi-relational environment. Moreover, we extend the benchmarking of our prototype implementation which is capable of distributing data (more precisely, vertical fragments) among several database servers and which appropriately analyzes and rewrites arbitrary SQL queries. In detail, in this article we make the following contributions:

– We investigate separation of duties in a realistic data model; in particular, we consider multiple database relations as well as dependencies between data which are allowed to span multiple relations – both of these are important in real-world database systems to achieve non-redundancy and to avoid anomalies for example by database normalization.
– We formulate separation of duties as a mathematical optimization problem according to several optimization goals – satisfying confidentiality constraints under dependencies while minimizing the amount of database servers, maximizing the amount of satisfied visibility constraints.
– We provide article a benchmarking of the optimization procedure (using IBM CPLEX) on the widely used TPC-E data schema.
– We give a detailed account of the query rewriting approach followed when accessing data that are distributed among different fragments.
– We provide details of a benchmarking of distributed query execution on the fragmented TPC-H dataset. In particular, our approach can execute all of the TPC-H benchmark queries – while currently existing approaches using property-preserving encryption are unable to execute the entire query set.

We start this article with a survey of related work in Sect. 2. Section 3 sets the necessary terminology; Sect. 4 analyzes the theory of several Separation of Duties problems; Sect. 5 provides a translation into an integer linear program; Sects. 6 and 7 describe the implementation and evaluation; Sect. 8 concludes the article.

2   Related Work

Security and privacy have been challenging tasks ever since the introduction of the principle of data outsourcing and a lot of research has been carried out to address different aspects of this problem such as access control, data confidentiality and data integrity. Security can be achieved by encryption – carried out by the user before outsourcing the data to the database. Still, encryption operations are generally very costly and moreover, existing cryptographic techniques can still not be efficiently used to evaluate more complex queries like, for instance, computations on the data. In this paper we focus on data confidentiality by data fragmentation and distribution without encryption; however our proposed approach can indeed be combined with conventional encryption (as in [1]) or even novel encryption approaches (that we also applied in prior work as for example order-preserving encryption [32], searchable encryption [33] or even fuzzy searchable encryption [21,23]).

Vertical fragmentation approaches split a database into two or more fragments consisting each of a subset of the attributes (that is, columns) of the entire database. The theory of vertical fragmentation for relational database systems is well-studied. As an early resource, [9] study vertical fragmentation that also takes transaction processing costs into account. Fragmentation is also extensively covered in the standard textbook [26]. A more recent comparative evaluation of vertical fragmentation approaches is provided in [25]; however all of these approaches do not consider fragmentation as a security mechanism.

Vertical fragmentation can be used to break sensitive associations between attributes. Vertical fragmentation for data outsourcing was first analyzed from a security point of view in [1] where a single relation is divided into *two* fragments. They model sensitive associations between columns of the single relation as subsets of its columns. In each fragment some values of a tuple are stored as plaintext while each fragment always additionally contains each entire tuple in encoded form (for example, encrypted); that is, they rely on encryption whenever it is impossible to meet the confidentiality requirements with those two servers only. The two fragments are outsourced to two distinct *honest-but-curious, non-communicating* servers. A user runs a client software for database management, query optimization and query postprocessing. Query execution performance is good for queries covering plaintext values but worse for querying encoded data. The authors provide strategies to optimize query execution plans and identify an optimal vertical fragmentation according to some cost matrix. In contrast to this work, our approach supports more than two external servers as well as a trusted owner server, so that we can refrain from using any form of encryption and we consider more optimization goals.

Several approaches extended and improved this first analysis. [12,14] pioneer the idea of a single data owner safeguarding as few data as possible of a single relation. To do so, a database table is split vertically into a *server fragment* (to be stored at an untrusted cloud server) and an *owner fragment* (to be stored at a trusted local database server). Again, only two fragments are considered. The aim is to store a minimal subset of columns in the local database such that the remaining set of columns which is stored at the untrusted server is insensitive. This approach works totally without encryption because the owner site is assumed to be trusted. The authors define fragmentation correctness with respect to completeness, confidentiality and non-redundancy, and analyze fragmentation metrics to assess quality of a fragmentation. A complexity analysis is based on the weighted minimum target hitting set problem.

Other proposals such as [10,11,13,15,17–19] use vertical fragmentation as a tool to split sensitive associations in a database table. It is required that those fragments do not have an attribute in common such that the fragments are *unlinkable*. Due to the unlinkability, the fragments can possibly be stored at a single untrusted server which is then in possession of all the data but cannot establish sensitive associations. In particular, [19] introduces the concept of k-loose associations: while associations between single tuples (or the confidential values therein) in two fragments remain protected, more general associations between groups (of size k) of confidential values can be published - and hence improve data visibility. For a detailed security analysis, probabilities of one-to-one associations between values given the published fragments and k-loose associations are analyzed. The authors consider both confidentiality and visibility constraints but assume that there is no conflict between these constraints.

In [18] the concept of *data dependencies* is introduced. The authors state that certain combinations of attributes can be used by a sophisticated untrusted server to draw conclusion about other attributes which could potentially lead to the exposure of sensitive data. More specifically, [2,4] consider certain classes of data dependencies (in particular, subcases of equality generating dependencies and tuple generating dependencies) and investigate their impact on information disclosure.

[8] proposes vertical fragmentation into two fragments: one with confidential and one with non-confidential attributes; only confidential attributes are encrypted with the Advanced Encryption Standard (AES). Associations between attributes are not considered. An analysis reveals that joins between the confidential and non-confidential attributes are costly.

We extend and consolidate these prior approaches by supporting multiple relations while combining several optimization goals into one problem definition.

In addition, some approaches go beyond vertical fragmentation and consider other kinds of fragmentations: Expressive constraints and dependencies in first-order logic have previously been analyzed in [4] for vertical as well as in [35] for horizontal confidentiality-preserving fragmentations.

Several approaches cover only data publishing – as opposed to data storage or data sharing as we do. In particular, [37,39] studies data publishing and hierarchical data partitioning under the notion of differential privacy which is a probabilistic measure of data confidentiality. Roughly, the probability distribution of the published data should not diverge substantially from similar data sets. To achieve this they partition the original datasets into subsets and introduce noise to ensure confidentiality for certain statistical queries; these queries are however not general enough (mostly count queries are considered). As another approach for data publishing is [24] that enumerates all possible hybrid fragmentations (what the authors call "hierarchical partitioning") of a table. Afterwards, an optimal fragmentation is anonymized by generalizing data in each fragment with a new value. More recently, [38] combine vertical fragmentation with k-anonymity, l-diversity, and t-closeness to achieve more efficiency when publishing multi-dimensional data. In contrast to these (as well as related) approaches, we focus on data storage and data sharing instead of data publishing – hence in our approach data are not distorted.

3   Background

As the underlying data model, we focus on the formal definition of a relational database. As usual, a relation schema $R(\{a_1, \ldots, a_n\})$, or simply $R(a_1, \ldots, a_n)$, consists of a *relation name* R and a finite set of *attributes* $\{a_1, \ldots, a_n\}$ with

$n \geq 1$. Each attribute a_i is associated with a specific domain which is denoted by the expression $\mathrm{dom}(a_i)$. Next, a *relation* r, also denoted by $r(R)$, over the relation schema $R(a_1, \ldots, a_n)$ is defined as an ordered set of *n-tuples* $r = (t_1, \ldots, t_m)$ such that each *tuple* r_j is an ordered list $t_j = v_1, \ldots, v_n$ of values $v_i \in \mathrm{dom}(a_i)$ or $v_i = \mathsf{NULL}$. The *degree* of a relation is defined as the number of attributes in r.

A *database schema* $D = \{R_1(A_1), \ldots, R_N(A_N)\}$ is defined by a name D and a set of relation schemes $R_i(A_i)$ where A_i denotes the corresponding set of attributes. Finally, a *database state* $d = \{r_1, \ldots, r_N\}$ over a database schema $D = \{R_1(A_1), \ldots, R_N(A_N)\}$ is a set of relations such that each r_i is a relation over the respective relation schema $R_i(A_i)$.

As a tuple identifier tid_s for relation schema R_s a subset of A_s is chosen that is a candidate key: the tuple values are functionally dependent on the key; that is, two tuples with the same identifier value would be identical on all other values. Formally, if r_s is a relation over the schema $R_s(A_s)$ and $\mathrm{tid}_s \subseteq A_s$, for the two tuples $t_1, t_2 \in r_s$ the following holds

$$t_1[\mathrm{tid}_s] = t_2[\mathrm{tid}_s] \Rightarrow t_1 = t_2$$

The set of all tuple identifiers is denoted by $\mathrm{tid} := \bigcup_{s=1}^{N} \mathrm{tid}_s$.

To illustrate the individual steps (namely, setup, select, insert, delete and update), a small database D consisting of two tables **D** and **P** in a hospital scenario serves as a running example. The first table stores information about doctors and the second table stores information about patients:

$$D = \{ \qquad \mathbf{D}(\mathsf{DocID}, \mathsf{Name}, \mathsf{DoB}, \mathsf{ZIP}, \mathsf{Specialty}),$$
$$\mathbf{P}(\mathsf{PatID}, \mathsf{Name}, \mathsf{DoB}, \mathsf{ZIP}, \mathsf{Diagnosis}, \mathsf{Treatment}, \mathsf{DocID}) \}$$

where DocID and PatID serve as tuple identifiers.

4 Separation of Duties Problems

Analogous to [1,12] we assume that Cloud service providers are "honest but curious". This means that servers handle requests and answer queries correctly; but, while they do not manipulate the stored or returned data, still they analyze data and user behavior and try to gain sensitive information from it.

A security policy consisting of confidentiality constraints (see Definition 1) describes what information is confidential in terms of subsets of attributes (that is, column names) of relations. The presented separation of duties approach aims at protecting confidentiality of either all values in an individual column (the so-called singleton constraints) or the combination of values for the same tuple in different columns (the so-called association constraints). As an extension to prior work, in a database containing multiple relations, sensitive associations can exist among relations. This is expressed in the following definition of multi-relational confidentiality constraints (Definition 1).

Definition 1 *(Multi-relational Confidentiality Constraints). Let the sets $A_1, \ldots, A_N$ denote sets of attributes that are pairwise disjoint; furthermore, let $D = \{R_1(A_1), \ldots, R_N(A_N)\}$ be a database schema and $d = \{r_1, \ldots, r_N\}$ a database state over D. A multi-relational confidentiality constraint on D is defined by a subset of attributes $c \subseteq \bigcup_{s=1}^{N} A_s$. A multi-relational confidentiality constraint c with $|c| = 1$ is called a* singleton constraint. *If $|c| > 1$ it is called an* association constraint.

The condition that the set of attributes are pairwise disjoint is introduced to assure that attributes can uniquely be associated with the according relation schema. This can easily be achieved by choosing a suitable naming convention (like prepending relation names to the attributes).

Vertical fragmentation is used to meet the security requirements expressed by the confidentiality constraints. In Definition 2, correctness of vertical fragmentation is defined based on the three properties of completeness, disjointness and reconstruction.

Definition 2 *(Multi-relational Vertical Fragmentation, cardinality of a fragmentation, physical fragments). Let $A_1, \ldots, A_N$ denote sets of attributes that are pairwise disjoint; furthermore, let $d = \{r_1, \ldots, r_N\}$ be a database state over the database schema $D = \{R_1(A_1), \ldots, R_N(A_N)\}$. A set of fragments $\mathbf{f} = (f_0, \ldots, f_k)$ where $f_j \subseteq \bigcup_{s=1}^{N} A_s$ for all f_j is called a* correct *(multi-relational) vertical fragmentation of d if the following conditions are met:*

- ***Completeness:*** *$\bigcup_{i=0}^{k} f_j = \bigcup_{s=1}^{N} A_s$*
- ***Disjointness:*** *$f_i \cap f_j \subseteq tid, \forall\, f_i \neq f_j$ with $f_i, f_j \neq \emptyset$*
- ***Reconstruction:*** *$tid_s \subset (f_j \cap A_s)$, if $f_j \cap A_s \neq \emptyset$*

A fragmentation that satisfies all properties except the disjointness is called a lossless *(multi-relational) vertical fragmentation of r. The* cardinality *of a vertical fragmentation $\mathrm{Card}(\mathbf{f})$ is defined as the number of nonempty fragments in $\mathbf{f}$:* $\mathrm{Card}(\mathbf{f}) = \sum_{\substack{j=0 \\ f_j \neq \emptyset}}^{k} 1$.

For fragment f_j, its multirelational *physical fragment is the set of projections*

$$d_j := \{\pi_{f_j \cap A_1}(r_1), \ldots, \pi_{f_j \cap A_N}(r_N)\}$$

which is a database over the database schema $D_j = \{R_1(f_j \cap A_1), \ldots, R_N(f_j \cap A_N)\}$. The individual relations $\pi_{f_j \cap A_s}(r_s)$ for $s \in \{1, \ldots, N\}$ are called relation fragments.

Note that each fragment f_j contains a subset of the attributes of each original table; the physical fragment d_j contains subtables of all the original tables. The cardinality of the fragmentation $\mathbf{f}$ denotes the amount of non-empty fragments f_j. We will later on minimize the cardinality in order to minimize the amount of occupied external cloud servers.

In this definition the completeness property ensures that every attribute is contained in at least one fragment. The disjointness property prevents unnecessary copies of attributes that are not tuple identifiers; in other words, only tuple identifiers are allowed to be contained in more than one fragment. The reconstruction property makes sure that a fragment contains all necessary tuple identifiers to reconstruct the original relations by joining the corresponding relation fragments. More precisely, the reconstruction property ensures that a tuple identifier tid_s is contained in a fragment f_j, if and only if f_j also contains a non-tuple identifier attribute from relation r_s. On the one hand, this ensures that the individual relations can be reconstructed using join operation on the tuple identifiers and on the other hand prevents fragments that only contain the tuple identifier but no non-tuple identifier attribute of a specific relation.

In the following subsections we discuss three variants of separation of duties by vertical fragmentation.

4.1 Standard Separation of Duties

As a minimum, confidentiality-preserving fragmentations must obey security policies by ensuring that not all attributes contained in a confidentiality constraint are part of a fragment at the same time. The only exception is the owner fragment f_0 that is stored on the trusted client side and contains all singleton constraints as well as subsets of association constraints if they cannot be satisfied by distributing their attributes among several server fragments.

Definition 3 (Confidentiality). *Let $A_1, \ldots, A_N$ denote pairwise disjoint sets of attributes. Given a database state $d = \{r_1, \ldots, r_N\}$ over the database schema $D = \{R_1(A_1), \ldots, R_N(A_N)\}$ and a set of confidentiality constraints C. A vertical fragmentation $\mathbf{f} = (f_0, \ldots, f_k)$ is* confidentiality-preserving *with respect to a set of confidentiality constraints C if the following condition is met:*

$$c \nsubseteq f_j \text{ for all } c \in C \text{ and } j \in \{1, \ldots, k\}$$

In this definition f_0 is the owner fragment (to be stored at the trusted database server) and $f_1, \ldots, f_k$ are the k server fragments (to be stored at the k untrusted database servers). A confidentiality-preserving vertical fragmentation therefore requires that the combination of attributes defined by a confidentiality constraint is *not* jointly visible in a server fragment. For singleton constraints, this implies that the corresponding attribute must be placed in the owner fragment.

To avoid redundancy and unwanted interactions with the tuple identifiers, we impose some restrictions on the set of confidentiality constraints. In particular, tuple identifiers are assumed to be unsensitive information – because they are needed to reconstruct the original relations – and hence should not be contained in confidentiality constraints.

Definition 4 (Well-defined Confidentiality Constraints). *Let $A_1, \ldots, A_N$ denote pairwise disjoint sets of attributes and let $tid_s \subset A_s$ denote the designated tuple identifier for $A_1, \ldots, A_N$ respectively. Moreover, let $d = \{r_1, \ldots, r_N\}$ denote a database state over the database schema $D = \{R_1(A_1), \ldots, R_N(A_N)\}$. A set of confidentiality constraints C is* well-defined *if it satisfies the following conditions:*

- $c \not\subseteq c'$ for all $c, c' \in C$ with $c \neq c'$
- $c \cap tid_s = \emptyset$ for all $c \in C$ and $s \in \{1, \ldots, N\}$ with $c \subseteq A_s$

The first condition requires that no confidentiality constraint c is a subset of another c' – due to the requirements for a confidentiality-preserving multi-relational vertical fragmentation, the restriction that c is not jointly visible in any server fragment already implies that c' is not jointly visible in any server fragment.

The second condition of the previous definition further implies that confidentiality constraints c that contain only attributes from a single relation are not allowed to contain tuple identifier attributes if they contain at least one non-tuple identifier attribute. Such constraints can simply be replaced by the semantically equivalent constraints $c \setminus tid_s$. This will avoid unnecessary case differentiations in the remainder of this work.

Continuing our example, confidentiality constraints can express that patients' and doctors' names are highly confidential and that (as a kind of quasi-identifiers) the combinations of a patient's DoB, ZIP code and diagnosis as well as a doctor's DoB and ZIP code must not be revealed together.

$$C = \{\{\mathbf{P}.\text{Name}\}, \{\mathbf{D}.\text{Name}\}, \{\mathbf{P}.\text{DoB}, \mathbf{P}.\text{ZIP}, \mathbf{P}.\text{Diagnosis}\}, \{\mathbf{D}.\text{DoB}, \mathbf{D}.\text{ZIP}\}\}$$

A confidentiality-preserving fragmentation consists of one owner fragment

$$f_0 = \{\mathbf{P}_0(\text{PatID}, \text{Name}), \mathbf{D}_0(\text{DocID}, \text{Name})\}$$

and two server fragments

$$f_1 = \{\mathbf{P}_1(\text{PatID}, \text{DoB}, \text{DocID}), \mathbf{D}_1(\text{DocID}, \text{ZIP})\}$$

and

$$f_2 = \{\mathbf{P}_2(\text{PatID}, \text{ZIP}, \text{Diagnosis}, \text{Treatment}), \mathbf{D}_2(\text{DocID}, \text{DoB}, \text{Specialty})\}.$$

It can be seen that no server fragment contains the entire set of attributes specified in one confidentiality constraint. Moreover, the tuple identifiers are the necessary information that enables the owner to reconstruct the two original tables. Again note that each fragment contains a subset of the attributes of each of the two original relations $\mathbf{P}$ and $\mathbf{D}$; the cardinality of our fragmentation is 3 because we obtained one owner fragment and two server fragments.

As a last component, we consider storage space capacities for the servers. We specify a weight function that assigns a weight to each set of attributes that denotes the capacity consumption of the set: $w_d : \mathcal{P}(A) \longrightarrow \mathbb{R}^+$. A simple weight function could for example count the number of attributes in the set. We then consider a maximum capacity W_j for each server S_j and require that the summed weights of the fragment do not exceed the capacity of the server that hosts this fragment.

With these preliminaries the definition of the Standard Multi-relational Separation of Duties Problem considering the cardinality of the fragmentation (as in Definition 2) as well as the confidentiality (as in Definition 3) is as follows:

Definition 5 *(Multi-relational Separation of Duties). Given a database schema $D = \{R_1(A_1), \ldots, R_N(A_N)\}$, a database state $d = \{r_1, \ldots, r_N\}$, a well-defined set of multi-relational confidentiality constraints C, tuple identifiers $tid_s \subset A_s$ for all $s \in \{1, \ldots, N\}$, a weight function w_d, servers $S_0, \ldots, S_k$ and corresponding maximum capacities $W_0, \ldots, W_k \in \mathbb{R}_0^+$, the Multi-relational Separation of Duties Problem consists of finding a correct confidentiality-preserving fragmentation $\mathbf{f} = (f_0, \ldots, f_k)$ of minimal cardinality $\mathrm{Card}(\mathbf{f})$ such that the capacities of the storage are not exceeded, i.e. $w_d(f_j) \leq W_j$ for all $0 \leq j \leq k$.*

The maximum capacity W_0 of the owner fragment can be set such that the owner fragment only stores the attributes in the singleton constraints – which cannot be outsourced due to their sensitivity. This can be achieved by choosing a suitable capacity of the owner fragment. Yet, it must be considered that the correct tuple-identifier attributes must also be part of the owner fragment to satisfy the reconstruction property.

Lemma 1. *If the set of singleton constraints $A^* := \{c \in C \mid |c| = 1\}$ denotes the set of all sensitive attributes, when setting W_0 to*

$$W_0 = \sum_{s:A_s \cap A^* \neq \emptyset} w_d(tid_s) + \sum_{c \in C:|c|=1} w_d(c).$$

the owner fragment only stores the attributes contained in singleton constraints and the necessary tuple identifiers.

The Standard Separation of Duties Problem can be viewed as a combination of two famous NP-hard problems, the bin packing problem due to the capacity constraints of the storage locations and the vertex coloring problem due to the confidentiality constraints.

4.2 Visibility Constraints

In the multi-relational scenario it is very important to control the resulting fragmentation in order to increase the utility of the fragmented database and avoid unnecessary joins when executing queries on the distributed fragments. Increased usability means that certain combinations of attributes are stored on a single server because they are often queried together. The notion of visibility constraints will be adapted to this scenario: visibility constraints are defined as subsets of attributes that should be placed in a single fragment – in this case, we say that the visibility constraint is satisfied. Satisfaction of visibility should only be enforced if the resulting fragmentation is not in conflict with the confidentiality requirements. That is, confidentiality requirements are ranked higher than visibility requirements. Formally, the definition of visibility constraints and the amount of satisfied visibility constraints is as follows:

Definition 6 *(Visibility constraint, satisfaction). Let $d = \{r_1, \ldots, r_N\}$ be a database state over the database schema $D = \{R_1(A_1), \ldots, R_N(A_N)\}$.*

A *(multi-relational)* visibility constraint *over D is a subset of attributes $v \subseteq$*
$\bigcup_{s=1}^{N} A_s$. *A multi-relational fragmentation* $\mathbf{f} = (f_0, \ldots, f_k)$ *satisfies v if there is*
a $0 \leq j \leq k$ *such that* $v \subseteq f_j$. *If such an* f_j *exists, define* $\mathrm{Sat}_v(\mathbf{f}) := 1$. *Otherwise, define* $\mathrm{Sat}_v(\mathbf{f}) := 0$. *For any set V of visibility constraints, the number of*
satisfied visibility constraints is

$$\mathrm{Sat}_V(\mathbf{f}) := \sum_{v \in V} \mathrm{Sat}_v(\mathbf{f})$$

Note that as opposed to other approaches we treat visibility constraints as
soft constraints; that is, conflicts in the specification are allowed and visibility
constraints will be satisfied only if the confidentiality can still be ensured. Hence,
it may happen that not all visibility constraints can be satisfied.

For example, focusing on our patient example table, in order to preserve the
privacy of the patients, the confidentiality constraint $c = \{\mathsf{DoB}, \mathsf{ZIP}\}$ is enforced;
in addition, a visibility constraint $v = \{\mathsf{ZIP}, \mathsf{Diagnosis}\}$ is introduced that enables
the statistical evaluation of the frequency of illnesses per ZIP code. Hence, one
possible privacy-preserving fragmentation is given by

$$\mathbf{f} = \{f_0, f_1, f_2, f_3\}$$

with:

$$f_0 = \emptyset, \qquad\qquad f_1 = \{\mathsf{PatID}, \mathsf{DoB}\},$$
$$f_2 = \{\mathsf{PatID}, \mathsf{ZIP}, \mathsf{Treatment}\}, f_3 = \{\mathsf{PatID}, \mathsf{Diagnosis}\}.$$

Another privacy-preserving fragmentation is given by

$$\mathbf{f}' = \{f_0', f_1', f_2', f_3'\}$$

with:

$$f_0' = \emptyset, \qquad\qquad f_1' = \{\mathsf{PatID}, \mathsf{DoB}\},$$
$$f_2' = \{\mathsf{PatID}, \mathsf{ZIP}, \mathsf{Diagnosis}\}, f_3' = \{\mathsf{PatID}, \mathsf{Treatment}\}.$$

The important thing to notice here is that both fragmentations satisfy the
confidentiality constraint but in $\mathbf{f}$ the attributes in v are spread among two
servers while in $\mathbf{f}'$ they are on the same server; more formally, $\mathrm{Sat}_v(\mathbf{f}) = 0$ while
$\mathrm{Sat}_v(\mathbf{f}') = 1$. As a result, the second fragmentation $\mathbf{f}'$ is better because a query
for the two attributes ZIP and $\mathsf{Diagnosis}$ can be answered by a single server (the
one hosting f_2') without the need to join on patient ID.

Minimizing the number of servers versus maximizing the number of fulfilled
visibility constraints are two contrary goals. That is why in the following definition of the Extended Multi-relational Separation of Duties problem we introduce a weighted sum of these two goals using two weights α_1 and α_2. Note that
satisfying the confidentiality constraints is still a hard constraint and will be
mandatory. Moreover, omitting the disjointness property of fragmentation helps
increase the number of fulfilled visibility constraints. Therefore, in the following
problem statement, only a *lossless* fragmentation is required.

Definition 7 (Extended Multi-relational Separation of Duties). *Given database schema* $D = \{R_1(A_1), \ldots, R_N(A_N)\}$, *a database state* $d = \{r_1, \ldots, r_N\}$, *a set of well-defined multi-relational confidentiality constraints* C, *visibility constraints* V, *tuple identifiers* $tid_s \subseteq A_s$ *for all* $s \in \{1, \ldots, N\}$, *a weight function* w_d, *servers* $S_0, \ldots, S_k$ *with maximum capacities* $W_0, \ldots, W_k \in \mathbb{R}_0^+$ *and positive weights* $\alpha_1, \alpha_2 \in \mathbb{R}_0^+$, *find a lossless confidentiality-preserving fragmentation* $\mathbf{f} = (f_0, \ldots, f_k)$ *of minimal cardinality which satisfies* $w_d(f_j) \leq W_j$ *for all* $0 \leq j \leq k$ *such that the weighted sum* $\alpha_1 \operatorname{Card}(\mathbf{f}) - \alpha_2 \operatorname{Sat}_V(\mathbf{f})$ *is minimal.*

Using the weighted sum serves the following two purposes:

1. $\alpha_1 \operatorname{Card}(\mathbf{f})$ is responsible for minimizing the cardinality (amount of fragments) of the fragmentation. Hence we aim to use as few external servers as possible to store the server fragments.
2. By subtracting $\alpha_2 \operatorname{Sat}_V(\mathbf{f})$ each satisfied visibility constraint will lower the overall objective. Hence we aim to maximize the amount of satisfied visibility constraints.

The obvious question arises of how to appropriately choose the weights α_1 and α_2. In the following lemma we show that choosing the weights such that $\alpha_2|V| < \alpha_1$ results in assigning highest priority to the minimization of the cardinality of the fragmentation. Among those cardinality-minimizing fragmentations the number of satisfied visibility constraints should be maximal.

Lemma 2. *Consider weights* $\alpha_1 > 0$ *and* $\alpha_2 > 0$ *satisfying* $\alpha_2|V| < \alpha_1$. *If* $\mathbf{f}$ *is a solution to the Extended Multi-Relational Single-Relational Separation of Duties Problem and* $\mathbf{f}'$ *is a lossless confidentiality-preserving fragmentation that does not violate the capacity constraints* $w_d(f_j) \leq W_j$ *for all* $0 \leq j \leq k$, *the following statements hold:*

1. $\operatorname{Card}(\mathbf{f}') \geq \operatorname{Card}(\mathbf{f})$
2. If $\operatorname{Card}(\mathbf{f}') = \operatorname{Card}(\mathbf{f})$, *then* $\operatorname{Sat}_V(\mathbf{f}) \geq \operatorname{Sat}_V(\mathbf{f}')$

Proof. Let $\mathbf{f}, \mathbf{f}'$ and α_1 and α_2 be as stated in the lemma. First, Statement 1 is proven by contradiction: suppose $\operatorname{Card}(\mathbf{f}') < \operatorname{Card}(\mathbf{f})$ which is equivalent to

$$\operatorname{Card}(\mathbf{f}) - \operatorname{Card}(\mathbf{f}') \geq 1 \tag{1}$$

because both $\operatorname{Card}(\mathbf{f})$ and $\operatorname{Card}(\mathbf{f}')$ are positive integer values. Furthermore, because $\mathbf{f}$ is a solution to the Extended Multi-Relational Separation of Duties Problem, the following inequality holds:

$$\alpha_1 \operatorname{Card}(\mathbf{f}) - \alpha_2 \operatorname{Sat}_V(\mathbf{f}) \leq \alpha_1 \operatorname{Card}(\mathbf{f}') - \alpha_2 \operatorname{Sat}_V(\mathbf{f}') \tag{2}$$

At most $|V|$ visibility constraints can be satisfied, such that $0 \leq \alpha_2 \operatorname{Sat}_V(\mathbf{f}) \leq \alpha_2|V|$. Thus, because $\alpha_2 \operatorname{Sat}_V(\mathbf{f}') \geq 0$ the following inequality can be derived:

$$\alpha_1 \operatorname{Card}(\mathbf{f}) - \alpha_1 \operatorname{Card}(\mathbf{f}') \leq \alpha_2 \operatorname{Sat}_V(\mathbf{f}) - \alpha_2 \operatorname{Sat}_V(\mathbf{f}') \leq \alpha_2 \operatorname{Sat}_V(\mathbf{f}) \leq \alpha_2|V| \tag{3}$$

Together, Inequality 3 and the assumption that $\alpha_2|V| < \alpha_1$ in the lemma lead to the following inequality:

$$\alpha_1 \operatorname{Card}(\mathbf{f}) - \alpha_1 \operatorname{Card}(\mathbf{f}') < \alpha_1 \tag{4}$$

As α_1 is assumed to be greater than zero, the inequality

$$\operatorname{Card}(\mathbf{f}) - \operatorname{Card}(\mathbf{f}') < 1 \tag{5}$$

must be satisfied which contradicts Inequality 1.

Up next, Statement 2 is proven by contradiction. Hence, it is assumed that $\operatorname{Card}(\mathbf{f}') = \operatorname{Card}(\mathbf{f})$ and $\operatorname{Sat}_V(\mathbf{f}) < \operatorname{Sat}_V(\mathbf{f}')$.

The inequality

$$\alpha_1 \operatorname{Card}(\mathbf{f}) - \alpha_2 \operatorname{Sat}_V(\mathbf{f}) \leq \alpha_1 \operatorname{Card}(\mathbf{f}') - \alpha_2 \operatorname{Sat}_V(\mathbf{f}') \tag{6}$$

as above can now be simplified to

$$\alpha_2 \operatorname{Sat}_V(\mathbf{f}') \leq \alpha_2 \operatorname{Sat}_V(\mathbf{f}) \tag{7}$$

due to the assumption that the cardinalities of $\mathbf{f}$ and $\mathbf{f}'$ are equal. This gives us $\operatorname{Sat}_V(\mathbf{f}) \geq \operatorname{Sat}_V(\mathbf{f}')$ – a contradiction to the assumption that $\operatorname{Sat}_V(\mathbf{f}) < \operatorname{Sat}_V(\mathbf{f}')$.

Finally, it should be noted that as for the standard problem, it is often desirable that the owner fragment only consists of the attributes contained in singleton constraints and the respective tuple identifiers. To achieve this, one can again choose the weight of the owner fragment as explained in Lemma 1.

4.3 Dependencies

To model correlations between data, database dependencies can be specified. For example, in a medical setting a specific treatment might disclose the diagnosed disease. In a data publishing and also a data sharing application, such dependencies on the one hand enable *users* to infer more information from retrieved data; on the other hand, in a separation of duties setting, dependencies can enable a *server* to deduce much more information even though it only stores fragments of a confidentiality-preserving fragmentation: in the example, the specific disease can be inferred which however might be highly confidential information. Such inferences disclosing confidential information must be avoided and hence dependencies have to be considered when applying separation of duties.

De Capitani di Vimercati et al. [18] have explored the technique of fragmentation to ensure data confidentiality in presence of dependencies among columns. We will adopt their notion of dependencies that are specified as rules with a left-hand side (the premise) and a right-hand side (the consequence); both the premise and the consequence are sets of on column names. The intended semantics is that of a functional dependency: any combination of values for the premise uniquely discloses a combination of values for the consequence. Dependencies on the patient table could for example be $\mathsf{DoB}, \mathsf{ZIP} \rightsquigarrow \mathsf{Name}$ (that discloses the name of a patient from the date of birth and zip code) or Treatment $\rightsquigarrow$ Diagnosis (that discloses a diagnosis from the treatment). De Capitani di Vimercati et al. explore this problem only in a single-relational environment. For our application, the definitions and theories will be translated into a multi-relational context.

Definition 8 (Data Dependency). *A dependency δ over a database schema $D = \{R_1(A_1), \ldots, R_N(A_N)\}$ is an expression of the form $X \rightsquigarrow Y$, with $X, Y \subset \bigcup_{s=1}^{N} A_s$ and $X \cap Y = \emptyset$. The left hand side of a dependency δ is called the premise while the right hand side is called the consequence of δ. For simplicity, the notations δ.premise and δ.consequence (or δ.p and δ.c, for short) are used to denote the respective part of the dependency.*

A simple approach to make the information disclosed by dependencies visible in a fragment is adding the implied attributes to the fragment (this approach is called fragment and dependency composition in [18]):

Definition 9 (Dependency Composition). *For a given database schema $D = \{R_1(A_1), \ldots, R_N(A_N)\}$, a subset $f_j \subseteq \bigcup_{s=1}^{n} A_s$ of attributes and a set Δ of dependencies, the composition of f_j with dependency $\delta \in \Delta$ is the set of attributes:*

$$f_j \otimes \delta = \begin{cases} f_j \cup \delta.consequence, & if \delta.premise \subseteq f_j \\ f_j, & else \end{cases}$$

Next, we adopt the notion of closure of a set of attributes as in [18]; the closure is a superset that is immune to dependency composition:

Definition 10 (Closure). *Let $A_1, \ldots, A_N$ denote pairwise disjoint sets of attributes and let $d = \{r_1, \ldots r_N\}$ be a database state over the schema $D = \{R_1(A_1), \ldots, R_N(A_N)\}$. Moreover, let Δ denote a set of dependencies. For any subset $f \subseteq \bigcup_{s=1}^{N} A_s$ of attributes the closure with respect to δ is defined as the minimal set $\overline{f}$ which satisfies $f \subseteq \overline{f} \subseteq \bigcup_{s=1}^{N} A_s$ and for all $\delta \in \Delta$ it holds that $\overline{f} \otimes \delta = \overline{f}$. If the subset f satisfies $f = \overline{f}$ it is called closed.*

If $\mathbf{f} = (f_0, \ldots, f_k)$ denotes a lossless/correct fragmentation of d, the closure of that fragmentation with respect to Δ is defined as $\overline{\mathbf{f}} := (\overline{f}_0, \ldots, \overline{f}_k)$. A fragmentation for which every server fragment $f_j \in \{f_1, \ldots, f_k\}$ is closed is called a closed fragmentation.

It is generally not possible to find a *closed* correct fragmentation satifying the disjointness property. Hence, the following problem statement focuses on finding a closed lossless multi-relational fragmentation.

Definition 11 (Multi-relational Separation of Duties in Presence of Data Dependencies). *Given a database state $d = \{r_1, \ldots, r_N\}$ over the given database schema $D = \{R_1(A_1), \ldots, R_N(A_N)\}$, tuple identifiers $tid_s \subset A_s$ for all $s \in \{1, \ldots, N\}$, a well-defined set of multi-relational confidentiality constraints C, visibility constraints V, a set of dependencies Δ, a weight function*

w_d, servers $S_0, \ldots, S_k$ with maximum capacities $W_0, \ldots, W_k \in \mathbb{R}_0^+$ and positive weights $\alpha_1, \alpha_2 \in \mathbb{R}_0^+$, find a **closed** lossless confidentiality-preserving fragmentation $\mathbf{f} = (f_0, \ldots, f_k)$ of d which satisfies $w_d(f_j) \leq W_j$ for all $0 \leq j \leq k$ such that the weighted sum $\alpha_1 \operatorname{Card}(\mathbf{f}) - \alpha_2 \operatorname{Sat}_V(\mathbf{f})$ is minimized.

We now discuss the influence of dependencies on solvability of the Multi-relational Separation of Duties problem. One might wonder, whether the condition that the fragmentation is closed will prevent finding a solution when a non-closed confidentiality-preserving fragmentation exists that would solve the problem. As opposed to [18], in our problem statement this could be the case due to the capacity constraints. As every fragment f_j is a subset of its closure $\overline{f}_j$ it also holds that $w_d(f_j) \leq w_d(\overline{f}_j)$. For the Extended Multi-relational Separation of Duties Problem it was recommended to choose the capacity of the owner fragment such that it can only hold the attributes contained in singleton constraints and the respective tuple-identifier attributes (see Lemma 1). When dependencies are taken into account however, using this capacity could make the problem unsolvable because additional attributes should be stored in the owner fragment because they are sensitive on their own due to dependency. To illustrate this, a dependency $\delta \in \Delta$ with $|\delta.premise| = 1$ and $\delta.premise \neq c$ for all $c \in C$ is supposed. The premise contains a single attribute which is not contained in a singleton constraint. At first glance, it seems that this attribute is not sensitive on its own and therefore, the attribute in $\delta.premise$ will not be placed in the owner fragment when choosing the owner capacity W_0 as described in Lemma 1. Because the problem statement requires a closed fragmentation, the server fragment that contains the attribute in $\delta.premise$ needs to hold $\delta.consequence$, too. Obviously, a problem arises if there exists a confidentiality constraint $c \in C$ with $c \subseteq \delta.premise \cup \delta.consequence$ because the fragmentation obeying W_0 cannot be confidentiality-preserving and therefore, no solution exists. Such situations occur if the closure of an attribute is sensitive – in other words the attributes revealed by a single attribute due to dependencies violate a confidentiality constraint. Therefore, the actual set of sensitive attributes is given by the union of the two sets

$$A^* := \{c \in C \mid |c| = 1\}$$

and

$$A^{**} := \bigcup_{s=1}^{N} \left\{ a_i^s \in A_s \mid \exists c \in C \text{ with } c \subseteq \overline{\{a_i^s\}} \right\}.$$

There are two possible solutions for this. The first one is to introduce new confidentiality constraints $c = \{a_i^s\}$ for all $a_i^s \in A^{**}$ which is justified because those attributes can be regarded as sensitive attributes. The other solution is to increase the capacity of the owner fragment such that it holds all the attributes in A^* and A^{**} and the necessary tuple-identifier attributes to ensure the reconstruction property of the fragmentation. We chose the second solution and accomplished it by defining the capacity of the owner fragment as:

$$W_0 = \sum_{s: A_s \cap (A^* \cup A^{**}) \neq \emptyset} w_d(\operatorname{tid}_s) + \sum_{a \in (A^* \cup A^{**})} w_r(a),$$

With this setting, the owner fragment only stores the minimum amount of attributes necessary to ensure confidentiality.

5   ILP Formulation

In the following subsections, we discuss in detail how we translate the Multi-relational Separation of Duties Problem in Presence of Data Dependencies into an integer linear program (ILP) representation. All indicator variables z_v, u_{vj} (both for the visibility constraints), $p_{\delta j}$ (for the dependencies), y_j (for the fragments) and x_{ij}^s (for the attributes) are binary. Moreover, we establish the convention that $s \in \{1, \ldots, N\}, a_i^s \in A_s$ (the attributes), $a_{i'}^s \in \text{tid}_s$ (the tuple identifiers), $j \in \{0, \ldots, k\}, c \in C$ (the confidentiality constraints), $v \in V$ (the visibility constraints), $\delta \in \Delta$ (the dependencies).

The overall ILP that results in a confidentiality-preserving closed fragmentation (according to confidentiality constraints $c \in C$ and dependencies $\delta \in \Delta$) that at the same time occupies a minimum amount of servers and maximizes the amount of satisfied visibility constraints $v \in V$ is shown in Fig. 1.

$$\min \quad \alpha_1 \sum_{j=0}^{k} y_j - \alpha_2 \sum_{v \in V} z_v \qquad (8)$$

$$\text{s.t.} \quad \sum_{j=0}^{k} x_{ij}^s \geq 1, \qquad (9)$$

$$\sum_{a_i^s \in A_s^*} x_{ij}^s \leq x_{i'j}^s |A_s^*|, \qquad (10)$$

$$\sum_{a_i^s \in A_s^*} x_{ij}^s \geq x_{i'j}^s, \qquad (11)$$

$$\sum_{s=1}^{N} \sum_{a_i^s \in A_s} w_d(a_i^s) x_{ij}^s \leq W_j y_j, \qquad (12)$$

$$\sum_{a_i^s \in c} x_{ij}^s \leq |c| - 1, \qquad (13)$$

$$\sum_{a_i^s \in v} x_{ij}^s \geq u_{vj} |v|, \qquad (14)$$

$$\sum_{j=0}^{k} u_{vj} \geq z_v, \qquad (15)$$

$$\sum_{a_i^s \in \delta.p} x_{ij}^s \leq |\delta.p| - 1 + p_{\delta j}, \qquad (16)$$

$$\sum_{a_i^s \in \delta.p} x_{ij}^s \geq p_{\delta j} |\delta.p|, \qquad (17)$$

$$\sum_{a_i^s \in \delta.c} x_{ij}^s \geq p_{\delta j} |\delta.c|, \qquad (18)$$

Fig. 1. Integer linear program

A solution to the Separation of Duties Problem in Presence of Data Dependencies can be derived from a solution to the ILP by constructing the fragments according to the following rule. Attributes a_i^s for which the corresponding variable x_{ij}^s is equal to 1 in the ILP solution are contained in fragment j provided that the fragment j shall be non-empty (which is denoted by $y_j = 1$ in the ILP solution).

$$f_j := \begin{cases} \bigcup_{s=1}^{N} \{a_i^s \in A_s \mid x_{ij}^s = 1\}, & \text{if } y_j = 1 \\ \emptyset, & \text{else} \end{cases}$$

5.1 Translating Confidentiality

Indicator variables $x_{ij}^s \in \{0,1\}$ are used to express that attribute $a_i^s \in A_s$ from relation r_s is placed on server j. For every $s \in \{1,\ldots,k\}$ let $A_s^* := A_s \setminus \text{tid}$ denote the set of non-tuple identifiers of A_s. Binary variables $y_1,\ldots,y_k \in \{0,1\}$ are introduced which take a value of one if fragment f_j should be non-empty and a value of zero otherwise. In the objective function (8) the expression $\alpha_1 \sum_{j=0}^{k} y_j$ minimizes the cardinality of the fragmentation. Condition (9) ensures that every attribute is placed in at least one fragment satisfying the completeness property. Constraint (10) conditions that if there is a non-tuple identifier attribute contained in fragment f_j, i.e. the left hand side of the inequality is greater than one, then the right hand side must equal $|A_s^*|$ which is fulfilled if the variable $x_{i'j}^s$ for the tuple identifier attribute $a_{i'}^s \in \text{tid}_s$ also equals one and is hence also part of the fragment. The definition of the reconstruction property of multi-relational fragmentation requires that the tuple identifiers tid_s are proper subsets of the fragments; Constraint (11) takes care of this by allowing the variables $x_{i'j}^s$ for attribute $a_{i'}^s \in \text{tid}_s$ to equal one if at least one variable x_{ij}^s belonging to a non-tuple identifier attribute $a_i^s \in A_s^*$ equals one. Constraint (12) makes sure that the storage capacities are not exceeded and that y_j must take a value of one, whenever any attribute a_i^s is included in fragment f_j. Lastly Condition (13) is used to guarantee a confidentiality-preserving fragmentation, because at most $c-1$ variables x_{ij}^s for $a_i^s \in c$ can be equal to one.

5.2 Translating Visibility

Additional binary variables u_{vj} are introduced for every *visibility* constraint $v \in V$ and fragment $j \in \{1,\ldots,k\}$. These variables should only take a value of one if all attributes contained in v are placed in fragment f_j. Moreover, indicator variables z_v for all visibility constraints $v \in V$ are used to indicate whether there is at least one fragment that contains all attributes of visibility constraint $v \in V$. In the objective function (8) the summand $-\alpha_2 \sum_{v \in V} z_v$ maximizes the number of satisfied visibility constraints.

The constraints that ensure the proper treatment of the visibility constraints are given by Conditions (14) and (15). The former ensures that for every fragment f_j and every visibility constraint $v \in V$ variable u_{vj} can only be equal to one if the visibility constraint is satisfied in fragment f_j. The latter then makes sure that z_v, the indicator variable for visibility constraint $v \in V$, can only be equal to one if there is at least one u_{vj} for $j \in \{1,\ldots,k\}$ that equals one, i.e. visibility constraint v is satisfied one at least on server.

5.3   Translating Dependencies

It can easily be seen that a server fragment f_j equals its closure $\overline{f}_j$ if and only if for every data dependency $\delta \in \Delta$ one of the conditions $\delta.p \not\subseteq f_j$ or $\delta.p \cup \delta.c \subseteq f_j$ is true [18]. Hence, to check whether a server fragment f_j is closed, one first has to check for every data dependency $\delta \in \Delta$ if $\delta.p \in f_i$. In the ILP formulation we introduce indicator variables $p_{\delta j} \in \{0,1\}$ for each data dependency $\delta \in \Delta$ and server $j \in \{1,\ldots,k\}$ that should take a value of one if and only if all attributes in $\delta.p$ are stored in server fragment f_j. After that, we have to make sure that if a dependency premise is contained in a fragment the consequence must also be contained. Together, Constraints (16) and (17) ensure that $p_{\delta j}$ equals one if and only if all attributes in $\delta.p$ should be placed into the same server fragment f_j. The sum on the left hand side of Condition (16) is at most $|\delta.p|$. If this is the case, then $p_{\delta j}$ must equal one because otherwise the expression on the right hand side would be smaller. Hence, if all attributes in the premise of δ are part of fragment f_j variable $p_{\delta j}$ must take a value of one. Furthermore, Condition (17) achieves that variable $p_{\delta j}$ will be zero otherwise. If the left hand side of the equality is smaller than $|\delta.p|$, i.e. not all attributes in the premise of δ are contained in fragment f_j, then constraint can only be fulfilled if the right hand side equals zero or in other words $p_{\delta j}$ equals zero.

Finally, Constraint (18) requires that all attributes in $\delta.c$ are part of server fragment f_j if all attributes in $\delta.p$ are part of f_j: if $p_{\delta j}$ is equal to one the right hand side of the inequality takes the value $|\delta.c|$. In this case the constraint can only be fulfilled if the sum on the right side is also $|\delta.c|$ which means that x_{ij}^s equals one for all $a_i^s \in \delta$. On the other hand if $p_{\delta j}$ equals zero the condition is always fulfilled.

6   Implementation

There are the following entities involved in the system:

- **Untrusted Database Servers:** These servers store the server fragments and can process queries involving their respective fragment only. The individual physical fragments are organized in database tables.
- **Trusted Database Server:** This server stores the owner fragment and manages connections to the untrusted servers. Most common DBMSs provide means to include database tables stored at remote servers. In PostgreSQL for example, this can be realized with so-called *Foreign Data Wrappers* and MySQL provides the *FEDERATED Storage Engine*. This enables the later presented distributed database client to issue adequate high-level (possibly SQL) queries directly to the trusted database server instead of issuing sub-queries to each individual database server and then calculating the desired result. Instead, the built-in query processor of the trusted database server will decide how the query is actually optimized and executed.
- **Distributed Database Client:** The client acts as an additional layer between the database users and the database servers. It computes the fragmentation using an ILP solver, stores the metadata of the fragmentation

(i.e. at which server the columns are stored) and processes and rewrites user queries such that they are based on fragments instead of relations of the original database. The distributed database client can either access the database servers directly if a query exclusively involves columns of a single fragment or it can issue queries to the trusted database server which makes all other fragments stored at the untrusted databases servers visible. Those queries are then analyzed by the database system's query processor which decides how the query is internally executed by applying adequate optimization techniques. Finally, the results of the queries are transferred to the user.

The major advantage of the presented framework is that the only component that we had to implement is the distributed database client (available at [5]) that relies on the advanced query optimization techniques already provided by today's DBMS. The chosen programming language is JAVA and the implementation relies on the popular open source DBMS PostgreSQL. Therefore, the trusted database server and the untrusted database servers need to run a standard installation of PostgreSQL. To solve the ILPs the IBM ILOG CPLEX solver is used due to its comprehensive range of available solving strategies and several off-the-shelf optimizations. Lastly, to analyze and rewrite the users' queries, the distributed database client uses the open source project JSQLParser which is a SQL parser for JAVA. After solving the ILP using the CPLEX solver, the distributed database client continues with creating databases on the necessary database servers to store the respective physical fragments. In particular, a new database is set up at the trusted database server to store the owner fragment. Subsequently, the table fragments are set up and populated with the data from the original database. Using the foreign data-wrapper module postgres_fdw, the tables stored at the untrusted servers are made visible in the newly created database at the trusted server. Therefore, the database at the trusted database server contains each of the table fragments either as a local table if the fragment is part of the owner's physical fragment or as foreign table, otherwise. As a final step, the distributed database client sets up views in the database of the trusted database server using the local and foreign tables which correspond to the tables of the original database.

We now describe in detail the functionality that is offered by our system. To illustrate the individual steps, a small database consisting of two tables in a hospital scenario serves as a running example. The first table stores information about patients and the second table stores information about doctors working in the hospital (Tables 1 and 2):

Setup. For the setup, the owner has to set up the database to be fragmented at the trusted database server and specify the designated tuple identifier columns of the table as primary key columns. Furthermore, the owner has to provide a configuration file to tell the distributed database client where the database servers are located and how much space is available on each server; see Fig. 2 for an example with one owner server and three remote servers. Additional files can be created to specify the confidentiality constraints, visibility constraints and data dependencies (see Figs. 3, 4 and 5). In contrast to the configuration

```
Name,  Address,           Port,  Username, Password, Capacity, IsOwner
S0,    192.168.178.92,    5432,  postgres, postgres, 2.0,      owner;
S1,    192.168.178.82,    5432,  postgres, postgres, INF;
S2,    192.168.178.87,    5432,  postgres, postgres, INF;
S3,    192.168.178.88,    5432,  postgres, postgres, INF;
```

Fig. 2. Example configuration file

```
patient.P_Name;
doctor.D_Name;
patient.P_DoB, patient.P_ZIP, patient.P_Diagnosis;
doctor.D_DoB, doctor.D_ZIP;
```

Fig. 3. Example confidentiality constraints

```
patient.P_Diagnosis, patient.P_ZIP;
patient.P_Treatment, doctor.D_Specialty;
```

Fig. 4. Example visibility constraints

```
patient.P_DoB, patient.P_ZIP > patient.P_Name;
patient.P_Treatment > patient.P_Diagnosis;
patient.P_Diagnosis > patient.P_Treatment;
doctor.D_DoB, doctor.D_ZIP > doctor.D_Name;
```

Fig. 5. Example dependencies

Table 1. Table *patient* with primary key column *P_PatientID*

P_PatientID	P_Name	P_DoB	P_ZIP	P_Diagnosis	P_Treatment	P_DoctorID
1	J. Doe	07.01.1986	12345	Flu	Nose spray	1
2	W. Lee	12.08.1974	23456	Broken Leg	Gypsum	2
3	F. Jones	05.09.1963	23456	Asthma	Asthma inhaler	1
4	G. Miller	10.02.1982	12345	Cough	Cough syrup	1

Table 2. Table *doctor* with primary key column *D_DoctorID*

D_DoctorID	D_Name	D_DoB	D_ZIP	D_ Specialty
1	H. Bloggs	04.02.1971	34567	Respiratory
2	G. Douglas	27.07.1965	23456	Fraction

file, those files are optional. Subsequently, the owner instructs the distributed database client to set up the vertically fragmented database by solving an optimization problem with the CPLEX solver based on the provided input data. The weights of the specific columns are computed automatically and do not have to be provided by the owner. After solving the ILP using the CPLEX solver, the

Table 3. Owner Fragment

P_PatientID	P_Name
1	J. Doe
2	W. Lee
3	F. Jones
4	G. Miller

D_DoctorID	D_Name
1	H. Bloggs
2	G. Douglas

Table 4. Server fragment 1

P_PatientID	P_DoB	P_DoctorID
1	07.01.1986	1
2	12.08.1974	2
3	05.09.1963	1
4	10.02.1982	1

D_DoctorID	D_ZIP
1	34567
2	23456

Table 5. Server fragment 2

P_PatientID	P_ZIP	P_Diagnosis	P_Treatment
1	12345	Flu	Nose spray
2	23456	Broken Leg	Gypsum
3	23456	Asthma	Asthma inhaler
4	12345	Cough	Cough syrup

D_DoctorID	D_DoB	D_ Specialty
1	04.02.1971	Respiratory
2	27.07.1965	Fraction

distributed database client continues with creating databases on the necessary database servers to store the respective fragments. In particular, a new database is set up at the trusted database server to store the owner fragment. Subsequently, the tables are populated with the data from the original database. In the example, the table fragments shown in Table 3 are stored in the owner fragment at server S_0. Server S_1 stores the table fragments shown in Table 4. Lastly, Server S_2 stores the table fragments shown in Table 5. We implemented this functionality using the foreign data-wrapper module postgres_fdw which makes the server fragments available via the trusted database server as foreign tables. As a final step, the distributed database client sets up views in the database of the trusted database server using the local and foreign tables which correspond to the tables of the original database as shown in Fig. 6.

SELECT. The distributed database client provides two possible ways of querying the vertically fragmented database. The first possibility involves explicitly **rewriting** the users' queries using the JSQLParser such that they act on table fragments instead of the original tables. The second possibility is based on the created **views**.

The advantage of explicitly rewriting the users' queries is that the distributed database client can analyze the queries to make educated decisions which columns and table fragments are actually involved in the query and omit those that are not. In the provided implementation, for each table involved in

```
CREATE OR REPLACE VIEW doctor
AS
    SELECT s2_doctor_frag.d_doctorid  AS d_doctorid,
           s0_doctor_frag.d_name      AS d_name,
           s2_doctor_frag.d_dob       AS d_dob,
           s1_doctor_frag.d_zip       AS d_zip,
           s2_doctor_frag.d_specialty AS d_specialty
    FROM   s2_doctor_frag
           LEFT JOIN s0_doctor_frag USING(d_doctorid)
           LEFT JOIN s1_doctor_frag USING(d_doctorid);

CREATE OR replace VIEW patient
AS
    SELECT s2_patient_frag.p_patientid AS p_patientid,
           s0_patient_frag.p_name      AS p_name,
           s1_patient_frag.p_dob       AS p_dob,
           s2_patient_frag.p_zip       AS p_zip,
           s2_patient_frag.p_diagnosis AS p_diagnosis,
           s2_patient_frag.p_treatment AS p_treatment,
           s1_patient_frag.p_doctorid  AS p_doctorid
    FROM   s2_patient_frag
           LEFT JOIN s0_patient_frag USING(p_patientid)
           LEFT JOIN s1_patient_frag USING(p_patientid);
```

Fig. 6. View creation

```
SELECT  p_name, p_diagnosis, d_name AS attending_doctor
FROM patient, doctor
WHERE patient.p_doctorid = doctor.d_doctorid AND p_name = 'J. Doe';
```

Fig. 7. Original SELECT statement

the query, the distributed database client assesses which columns are needed and greedily chooses table fragments to obtain all necessary columns. As an example of a rewritten query, the SQL statement in Fig. 7 is rewritten inside the distributed database client into the SQL query shown in Fig. 8.

There are two major points to notice in this query. First, only servers S_0 and S_1 take part in the query, i.e. the fragments stored on S_3 are not affected. Furthermore, the WHERE condition is pushed down in to the SELECT queries affecting the table fragments as far as it can be processed by the respective server. This could potentially lead to less data being transferred during the execution process when tuples that do not satisfy the condition are excluded. This example also shows the major drawback of explicitly rewriting the queries which lies in the complexity of SQL that makes query rewriting a very challenging task.

In contrast, using views over the table fragments to recreate the original tables is a simple strategy to avoid query rewriting. The queries remain unchanged but logically they involve the views instead of real physical tables.

```
SELECT p_name, p_diagnosis, d_name AS attending_doctor
 FROM (SELECT
patientS0.p_name      AS p_name,
patientS0.p_patientid AS p_patientid,
patientS2.p_diagnosis AS p_diagnosis,
patientS1.p_doctorid  AS p_doctorid
  FROM  (SELECT p_name, p_patientid
   FROM   s0_patient_frag
   WHERE  ( true AND p_name = 'J. Doe' )) AS patientS0
   INNER JOIN (SELECT p_diagnosis, p_patientid
   FROM   s2_patient_frag
   WHERE  ( true AND true )) AS patientS2
   ON patientS0.p_patientid =
   patientS2.p_patientid
   INNER JOIN (SELECT p_doctorid, p_patientid
   FROM   s1_patient_frag
   WHERE  ( true AND true )) AS patientS1
   ON patientS2.p_patientid = patientS1.p_patientid) AS patient,
   (SELECT
   doctorS0.d_doctorid AS d_doctorid,
   doctorS0.d_name     AS d_name
   FROM  (SELECT d_doctorid, d_name
    FROM   s0_doctor_frag
    WHERE  ( true AND true )) AS doctorS0) AS doctor
    WHERE   patient.p_doctorid = doctor.d_doctorid AND p_name = 'J. Doe';
```

Fig. 8. Rewritten SELECT statement

```
INSERT INTO doctor
VALUES (3, 'C. Hall', '12/11/1990', 12345, 'Dermatology');
```

Fig. 9. Original INSERT statement

Therefore, it is up to PostgreSQL's query processor to decide how these queries are executed. While this method is easy to implement, its major drawback lies in the fact that each query involves all the columns of each table and unnecessary table fragments are not omitted from the query. For example, a query that selects only one column of a table fragmented among three servers will always involve all three of the servers although one would be sufficient. To the best of our knowledge, the PostgreSQL query processor does currently not consider excluding JOIN clauses of tables that are not affected by the query.

INSERT. When the distributed database client receives a request to insert a specific row into a table, it determines the affected table fragments and inserts a row, restricted to the according columns, in each of those. In the example, an INSERT statement of the form shown in Fig. 9 is translated into the three INSERT statements shown in Fig. 10.

```
INSERT INTO s0_doctor_frag (d_doctorid,d_name)
VALUES (3, 'C. Hall');

INSERT INTO s2_doctor_frag (d_doctorid, d_dob, d_specialty)
VALUES (3, '12/11/1990', 'Dermatology');

INSERT INTO s1_doctor_frag (d_doctorid,d_zip)
VALUES (3, 12345);
```

Fig. 10. Rewritten INSERT statement

```
DELETE FROM doctor WHERE d_name='G. Douglas';
```

Fig. 11. Original DELETE statement

DELETE. Deletion of rows can be broken down into two parts. Because table fragments only contain a subset of columns, it is generally not possible to evaluate the condition specified by the WHERE clause of the SQL DELETE statement with a single table fragment. Therefore, as the first part, a SELECT query is used to detect the values of the tuple identifiers of the rows that should be deleted. This SELECT query is executed by the distributed database client by either using the defined views or explicitly creating an according SELECT statement based on table fragments. The result of this query is stored in a temporary table. Subsequently, DELETE queries are executed for each table fragment with the condition that the tuple identifier values are present in the temporary table. Finally, the temporary table is deleted. In the prototype implementation, this is done by putting all of those queries into a single transaction, which is a sequence of SQL statements that is being executed consecutively and if specified, automatically deletes the created temporary tables at the end of the transaction. To illustrate this, the DELETE statement in Fig. 11 results in the transaction shown in Fig. 12. Note that the keyword TEMP specifies a temporal table and the ON COMMIT DROP option specifies that the temporal table is deleted at the end of the transaction.

UPDATE. Performing updates on rows of a specific vertically fragmented table resembles the deletion of rows due to the fact that the WHERE condition has to be specified. Therefore, as the first step, a temporal table is created that stores the tuple identifier columns of the affected rows. Subsequently, the distributed database client determines the involved table fragments and performs an update operation on each of those. To make sure that the proper rows are updated, the condition that the tuple identifier values are present in the temporal table is enforced. Those operations are again executed in a single transaction and the temporal table is dropped when the transaction is committed. Consider the UPDATE query in Fig. 13 for which the resulting transaction is shown in Fig. 14.

```
START TRANSACTION;
CREATE TEMP TABLE tmpdelete ON COMMIT DROP AS
 (SELECT d_doctorid
  FROM (SELECT doctors0.d_doctorid AS d_doctorid,
               doctors0.d_name     AS d_name
    FROM (SELECT d_doctorid, d_name
     FROM s0_doctor_frag
       WHERE  d_name = 'G. Douglas') AS doctors0) AS doctor
     WHERE  d_name = 'G. Douglas');

DELETE FROM s0_doctor_frag
WHERE (d_doctorid) IN (SELECT * FROM tmpdelete);

DELETE FROM s2_doctor_frag
WHERE (d_doctorid) IN (SELECT * FROM tmpdelete);

DELETE FROM s1_doctor_frag
WHERE (d_doctorid) IN (SELECT * FROM tmpdelete);

COMMIT;
```

Fig. 12. Rewritten DELETE statement

```
UPDATE patient SET p_zip = 23456 WHERE p_name = 'G. Miller';
```

Fig. 13. Original UPDATE statement

```
START TRANSACTION;
CREATE TEMP TABLE tmpupdate ON COMMIT DROP AS
 (SELECT p_patientid FROM
   (SELECT patientS0.p_patientid AS p_patientid,
           patients0.p_name      AS p_name
    FROM (SELECT p_patientid, p_name
      FROM s0_patient_frag WHERE p_name = 'G. Miller')
           AS patientS0) AS patient
     WHERE p_name = 'G. Miller');

UPDATE s2_patient_frag SET p_zi=23456
 WHERE (p_patientid)IN (SELECT * FROM tmpupdate);
COMMIT;
```

Fig. 14. Rewritten UPDATE statement

7 Evaluation

The prototype implementation is tested with two popular TPC benchmarks (TPC-E and TPC-H) for databases [29, 30]. Each of these benchmarks is suitable to evaluate different aspects of the Separation of Duties Problem. The database defined by the TPC-E benchmark consists of 33 tables and a total number of 191 columns; we only use the TPC-E schema which due to its size it is suitable to evaluate the effects of the number of constraints and dependencies on the performance of the ILP solver and the resulting fragmentations. In contrast, from the TPC-H benchmark we use data as well as queries in order to test the distributed runtime performance of our approach. It consists of a database containing 8 tables that models a typical database in a business environment. Moreover, it defines 22 complex SQL queries that are typical in decision support scenarios. Therefore, this benchmark is well-suited to test the implementation's capabilities in terms of query processing.

All of the tests were executed on a single PC equipped with an Intel Xeon E3-1231v3 @ 3.40 GHz (4 Cores), 32 GB DDR3 RAM and a Seagate ST2000DM001 2TB HDD with 7200 rpm. The PC is running Ubuntu 16.04 LTS. The database servers – including the trusted database server hosting the owner fragment – are running in separate, identical virtual machines which are assigned 4 cores and 8 GB of RAM. The virtual machines are running Ubuntu Server 16.04 LTS with an instance of PostgreSQL 9.6.1 installed. By running the servers in identical virtual machines, it is guaranteed that the results are not influenced by hardware or software differences. Lastly, the CPLEX version used by the implementation is CPLEX 12.7.

7.1   TPC-E Data Set

The TPC-E benchmark is intended to model the workload of a brokerage firm. It consists of 33 database tables which fall into four categories [29]:

- **Customer tables:** There are 9 tables that contain information about the brokerage firm's customers.
- **Broker tables:** There are 9 tables that contain information about the brokerage firm.
- **Market tables:** There are 11 tables that contain information about companies, markets, exchanges and industry sectors.
- **Dimension tables:** There are 4 tables that contain common information like zip codes or addresses.

Using the TPC-E database schema, our tests explore the influence of different sizes of sets of well-defined confidentiality constraints, visibility constraints and dependencies on the solver's performance and the resulting fragmentation. The trusted database server's capacity is set to zero: it should not store any data; this means that for the tests, singleton constraints are disallowed in the confidentiality constraints; moreover, the case that an association constraint contains an attribute that (after applying all possible dependencies) has an entire confidentiality constraint in its closure is disallowed, too: more formally, we disallow attributes a such that there is a constraint c with $c \subseteq \bar{a}$. This restriction is introduced to allow a maximal number of possible choices for the placement of the attributes during the optimization process because attributes contained in singleton constraints or single attributes with a sensitive closure can only be placed in the owner fragment which would limit the number of decisions the ILP solver has to draw.

Settings. The constraints and dependencies are generated randomly. To measure scalability of the approach, we introduce scale factors $\sigma_C, \sigma_V, \sigma_\Delta \in \mathbb{R}_0^+$ for confidentiality constraints, visibility constraints and dependencies, respectively. Note that the overall number of non-primary key columns is $n = 142$ in the TPC-E database. The scale factors can be interpreted as constraints/dependencies per non-primary key column. Because the primary-key columns act as tuple identifiers, they are supposed to be insensitive and are therefore neither part of the constraints nor part of the premise or consequence of dependencies.

Hence, if $A^* = \bigcup_{s=1}^{33} A_s \setminus \text{tid}_s$ denotes the set of all non-primary key columns of the 33 tables and $\mathcal{P}(A^*)$ denotes its powerset, the selection process is carried out as follows:

- **Confidentiality Constraints:** For each of the designated cardinalities $\nu \in \{2, 3, 4, 5\}$, sets of confidentiality constraints $C_\nu \subseteq \mathcal{P}(A^*)$ of cardinality $|C_\nu| = \lceil \frac{n}{4} \cdot \sigma_C \rceil$ are selected randomly from A^*. For each ν and C_ν it holds that $|c| = \nu$ for all $c \in C_\nu$. The resulting set of confidentiality constraints is then given by $C := \bigcup_{\nu=2}^{5} C_\nu$. Moreover, during the generation of the confidentiality constraints it is ensured that the resulting set C is well-defined. Therefore, this process results in a well-defined set of confidentiality constraints C of cardinality $|C| = 4 \cdot \lceil \frac{n}{4} \cdot \sigma_C \rceil$ which is equally divided into confidentiality constraints of cardinality $2, 3, 4$ and 5. A scale factor of $\sigma_C = 1$ would therefore result in a set of 144 confidentiality constraints which corresponds roughly to the size of non-primary key attributes in the database. As confidentiality constraints with lower cardinality are generally harder to satisfy than constraints with high cardinality, restricting ν to the values $\{2, 3, 4, 5\}$ is not a serious limitation.
- **Visibility Constraints:** Generating the visibility constraints is carried out similarly to the generation of confidentiality constraints. The only difference in the process is that the resulting set does not have a limitation of being well-defined. Hence, for each of the cardinalities $\nu \in \{2, 3, 4, 5\}$, sets of visibility constraints $V_\nu \subseteq \mathcal{P}(A^*)$ of cardinality $\lceil \frac{n}{4} \cdot \sigma_V \rceil$ are selected randomly such that $|v| = \nu$ for all $v \in V_\nu$. The overall set of visibility constraints is hence given by $V := \bigcup_{\nu=2}^{5} V_\nu$ which has a cardinality of $|V| = 4 \cdot \lceil \frac{n}{4} \cdot \sigma_V \rceil$ and is equally divided into visibility constraints of cardinality $2, 3, 4$ and 5.
- **Dependency:** Sampling dependencies is carried out differently because a dependency $\delta = \delta.premise \rightsquigarrow \delta.consequence$ is defined by the two sets $\delta.premise$ and $\delta.consequence$. The scale factor σ_Δ determines the cardinality of the set of dependencies Δ which is given by $|\Delta| = \lceil n \cdot \sigma_\Delta \rceil$. The dependencies itself are generated iteratively as follows: First, two random values $\nu_p \in \{2, \ldots 5\}$ and $\nu_v \in \{1, \ldots, 5\}$ are determined which define the cardinalities of the premise and the consequence. Then, $\delta.premise \subseteq \mathcal{P}(A^*)$ and $\delta.consequence \subseteq \mathcal{P}(A^*)$ are chosen randomly such that $|\delta.premise| = \nu_p$ and

$|\delta.consequence| = \nu_c$. This process is executed $\lceil n \cdot \sigma_\Delta \rceil$ times to obtain the final set of dependencies. The cardinality of the premise of each dependency is restricted to a value between 2 and 5 because on the one hand a premise of cardinality one could make a single attribute sensitive and on the other hand if the cardinality gets too large, it will become easier to place the attributes into multiple fragments such that the dependency only slightly influences the resulting fragmentation. Moreover, the cardinality of the consequence of each dependency is restricted to be smaller than 5 to guarantee a moderate balance between the cardinality of the premise of a dependency and the impact it has on the resulting fragmentation in terms of its consequence.

For each of the executed test runs the weights α_1 and α_2 required by the problem statement of the Multi-relational separation of duties problem in presence of data dependencies are chosen such that they satisfy the inequality $\alpha_2|V| < \alpha_1$ presented in Lemma 2. Therefore, the resulting fragmentation shall be of minimal cardinality and the number of satisfied visibility constraints shall be maximal among all feasible fragmentations of minimal cardinality.

We executed several test runs with these settings. For all test runs a time limit of 30 min is set for the optimization process. Previous tests have shown that after this time a feasible solution can be found but the objective value does not significantly improve after this time.

Furthermore, different measurements are introduced to measure the quality of the resulting fragmentation. These measurements are based on the *objective value* of the best integer solution obj_I found by CPLEX and the *lower bound* obj_{LP} on the objective value which could be established by CPLEX during the optimization progress by solving the LP-Relaxation of different subproblems of the ILP; the LP-Relaxation of an ILP is obtained by allowing the variables to take continuous instead of integral values. These measurements are defined as follows:

– **Relative MIP gap:** The *relative MIP gap* is a well-known general expression used by ILP solvers such as CPLEX to measure the quality of calculated solutions of mixed integer linear programs, i.e. linear programs of which some of the variables are restricted to be integer and others are real valued. Of course, it is also applicable for the special class of integer linear programs and it is defined by the following expression:

$$\frac{|obj_I - obj_{LP}|}{|obj_I|}$$

The *relative MIP gap* measures the percentage of how much the objective value of an optimal solution can maximally deviate from the objective value obj_I of a feasible solution due to the established lower bound. Therefore, if this measure equals p, there is an uncertainty whether the objective value could potentially be reduced by up to p percent. Our overall objective function considers both the minimization of the cardinality (number of external servers) as well as the satisfaction of the visibility constraints.

– **Card gap:** We introduce this measure specifically for the Separation of Duties Problem to account for the quality of a feasible solution's fragmentation and it is defined as follows:

$$\left\lfloor \frac{|obj_I - obj_{LP}|}{|\alpha_1|} \right\rfloor$$

The purpose of this expression is to measure the uncertainty about the fragmentation's cardinality of a feasible solution. For example, if this expression equals one, then it might be possible to reduce the fragmentation's cardinality by one and potentially, one database server less is necessary. If this expression equals zero, the fragmentation's cardinality is minimal and the number of servers necessary cannot be reduced.

– **Sat gap:** Similarly, we introduce this measure to account for the uncertainty of a solution in terms of visibility constraints with the following expression:

$$\left\lfloor \frac{|obj_I - obj_{LP}|}{|\alpha_2|} \right\rfloor$$

If this expression equals zero, the number of satisfied visibility constraints is maximal for the feasible solution. Further, if this expression equals $n \geq 1$, it is uncertain, whether up to n more visibility constraints could potentially be satisfied.

If all of these measures are equal to zero for a feasible solution's objective value obj_I and the established lower bound obj_{LP}, an *optimal* solution has been found. However, these expressions require a good feasible solution on the one hand, and a good lower bound on the other hand and if either of those cannot be found a high uncertainty remains.

Test Runs. To study the impact of confidentiality constraint, visibility constraints and dependencies individually, the evaluation is structured into several test cases.

First, the effects of increasing the number of confidentiality constraints is studied; hence, the scale factors σ_V and σ_Δ are set to zero (no visibility constraints and no dependencies) and different values for σ_C are tested (I). Next, σ_C is set to four, σ_Δ is set to zero and different scale factors σ_V are used to test the influence of an increasing number of visibility constraints (II). Lastly, to test the effects of the number of dependencies σ_C is set to four, σ_V is set to 0.25 and different scale factors σ_Δ are evaluated (III).

For the first test case (I), the σ_V and σ_Δ is set to zero and the number of confidentiality constraints is increased with the scale factor $\sigma_C \in \{1, 2, 4, 8, 16\}$. The results of these test runs are summarized by Table 6. An optimal solution is found nearly all of the scenarios as the *relative MIP gap* and the *Card gap* shows; that is, it is not possible to find a solution with less external servers. As the scale factor increases, more confidentiality constraints can only be satisfied when the individual original tables are split into more fragments (that is, the cardinality of the fragmentation increases). In the scenario with 2272 confidentiality constraints in four out of five cases we met our timeout limit of 1800 s and stopped the execution of CPLEX. Each of the fragments of one original table has to be stored on a server separate from the other servers storing fragments of the same original table; thus, the number of necessary database servers increases, too: for a scale factor of $\sigma_C = 1$ (that is, 144 confidentiality constraints), two servers are sufficient, for scale factors $\sigma_C = 2$ (284 confidentiality constraints) and $\sigma_C = 4$ (568 confidentiality constraints), three servers are necessary, for $\sigma_C = 8$ (1136 confidentiality constraints), there have to be four servers and for $\sigma_C = 16$ (2272 confidentiality constraints), five database servers have to be used. The runtime to find the optimal solution increases significantly for a scale factor of $\sigma_C = 16$; further optimizations of the solver could be employed to speed this setting up.

Table 6. Increasing number of confidentiality constraints (average over 5 runs)

σ_C	$|C|$ (# conf. constraints)	Cardinality (# servers)	MIP gap	Card gap	Optimal?	Time (s)
1	144	2	0%	0	Yes	0.61
2	284	3	0%	0	Yes	1.79
4	568	3	0%	0	Yes	1.55
8	1136	4	0%	0	Yes	11.99
16	2272	5	16%	0	In 1 of 5 runs	In 1 run 297.27 (otherwise timeout)

For the second test case (II), only the scale factor σ_V is changed and $\sigma_C = 4$ (568 confidentiality constraints) and $\sigma_\Delta = 0$ (no dependencies) remain fixed. In other words, the effects of increasing the number of visibility constraints are evaluated. For all of the runs, the resulting fragmentation has a cardinality of three (which is minimal); because the weights α_1 and α_2 have been chosen according to Lemma 2, the visibility constraints do not affect the cardinality of the fragmentation. The overall results of test runs are presented in Table 7. When the number of introduced visibility constraints increases, the percentage of satisfied constraints decreases (see column "Sat"). Note that not all visibility constraints can be satisfied because they are conflicting with confidentiality constraints. The column "Sat gap" tells us how many more visibility constraints could potentially be satisfied in an optimal solution.

The most important thing to notice is that only the scenario with the lowest number of visibility constraints ($\sigma_V = 0.25$ corresponds to 36 visibility constraints) can be solved optimally and for this scenario the time increases significantly compared to the same scenario without visibility constraints (see $\sigma_C = 4$ in Table 6). The other three scenarios (72, 144, 284 visibility constraints, respectively) exceeded the time limit and were canceled without having found an optimal solution in terms of number of fragments and satisfied visibility constraints. One way to improve the results could therefore be to develop provably good

Table 7. Increasing number of visibility constraints (average over 5 runs)

σ_V	$|V|$(#vis. constraints)	Sat	Cardinality (# servers)	MIP gap	Sat gap	Optimal?	Time (s)
0.25	36	27/36	3	0%	0	Yes	101.84
0.5	72	44/72	3	1.83%	7.4	No	Timeout
1	144	68/144	3	4.08%	33.6	No	Timeout
2	284	105/284	3	6.85%	113.6	No	Timeout

heuristics to provide good starting solutions for the solver on the one hand and on the other hand, to establish tight lower bounds to point the solver in the right direction and allow less choices for the variables. Moreover, what is also an important conclusion of these results is that visibility constraint should not be viewed as a means to allow the execution of as much queries as possible on a single server. Rather, they should be used selectively to speed up a small amount of queries that are particularly relevant for the database.

Finally for the third test case (III), the effects of increasing the number of data dependencies are analyzed (see Table 8). For that, the scale factors $\sigma_C = 4$ (568 confidentiality constraints) and $\sigma_V = 0.25$ (36 visibility constraints) are fixed and the scale factors $\sigma_\Delta \in \{1, 2, 4, 8, 16\}$ (corresponding to 142, 284, 568, 1136 and 2272 dependencies, respectively) are used for the data dependencies. Hence, these results resemble very much the scenario with $\sigma_C = 4$ and $\sigma_V = 0.25$ of the previous test runs with the additional introduction of data dependencies:

Table 8. Increasing number of data dependencies (average over 5 runs)

σ_Δ	$|\Delta|$ (# dependencies)	Card (# servers)	Sat	MIP gap	Optimal?	Time (s)
1	142	3	26.4/36	0%	Yes	60.22
2	284	3	27.2/36	0%	Yes	75.91
4	568	3	27/36	0%	Yes	41.36
8	1136	3	25/36	0%	Yes	47.18
16	2272	3	23.6/36	0%	Yes	21.57

All of the scenarios are solved optimally. A noticeable result is that increasing the number of data dependencies can in fact reduce the time needed to solve the problem. An important take-away message from these test runs is that the solver benefits from introducing data dependencies instead of using excessively many confidentiality constraints.

7.2   TPC-H Data Set

The TPC-H benchmark is described as a decision support benchmark. This means, that it simulates a system used to support decision making in business applications. We deemed this an appropriate setting to test distributed query execution on the vertically fragmented data set. The 8 tables in the TPC-H schema are $customer, part, partsup, supplier, lineitem, orders, customer, nation$ and $region$.

Unfortunately, the TPC-H data generator does not support PostgreSQL and therefore, other tools had to be used to set up the TPC-H benchmark with PostgreSQL. To set up the test database, the HammerDB [22] tool was used with a scale factor of 1. Moreover, the query generator provided by DBT-3 [16] was used to obtain the 22 TPC-H queries conforming with PostgreSQL's standard.

7.3 Settings

The number of tables in the TPC-H database is reasonably small, so that the following artificial scenario is used as the foundation for the tests:

- **Confidentiality Constraints:** The following rules are established for defining the constraints:
  - The name and the account balance of the customers and suppliers are sensitive:
    $c_1 = \{customer.c\_acctbal\}$,
    $c_2 = \{supplier.s\_acctbal\}$
  - The discount given on any order is sensitive:
    $c_3 = \{lineitem.l\_discount\}$
  - A customer's name and its address cannot be placed in the same server fragment:
    $c_4 = \{customer.c\_name, customer.c\_address\}$
  - A customer's name can not be associated with a specific order:
    $c_5 = \{customer.c\_name, orders.o\_custkey\}$
  - A supplier's name can not be associated with a line item:
    $c_6 = \{supplier.s\_name, lineitem.l\_suppkey\}$
  - The date of an order can not be associated with the total price:
    $c_7 = \{orders.o\_odate, orders.o\_totalprice\}$
  - A supplier's name can not be associated with the supplier's cost for a specific part:
    $c_8 = \{supplier.s\_name, partsupp.ps\_suppkey, partsupp.ps\_supplycost\}$
- **Dependencies:** Moreover, the following dependencies are introduced, concerning personal information about the customers and suppliers:

$$\delta_1 = \{customer.c\_address\} \rightsquigarrow \{customer.c\_name\}$$
$$\delta_2 = \{customer.c\_phone\} \rightsquigarrow \{customer.c\_name\}$$
$$\delta_3 = \{supplier.s\_address\} \rightsquigarrow \{supplier.s\_name\}$$
$$\delta_4 = \{supplier.s\_phone\} \rightsquigarrow \{supplier.s\_name\}$$

- **Visibility Constraints:** As the main purpose of visibility constraints is to speed up the execution of specific queries, a visibility constraint is introduced for each of the 22 queries consisting of all attributes in the query. Therefore, if a visibility constraint is satisfied, the execution of the corresponding query potentially involves a single database server only.

The weights α_1 and α_2 that are also needed for the problem statement are chosen to satisfy the inequality presented in Lemma 2. Finding an optimal solution to this specific instance of the Multi-relational Separation of Duties Problem and setting up the vertically fragmented database takes around two and a half minutes and the resulting fragmentation satisfies 5 of the 22 visibility constraints. Overall, the tables are distributed among 12 table fragments on a total of 3 database servers. One of those is the trusted database server and the remaining two are untrusted.

7.4   Test Runs

After the database is set up, the execution time of the 22 queries can be analyzed. For that, each query is executed with the following methods:

1. The original non-fragmented database is queried. To ensure the comparability, the original database is stored separately at the trusted database server.
2. The queries are rewritten by our trusted database client to act on table fragments instead of the original tables. We measured the time for executing (t), the time for rewriting the query (tr), the overall number of table fragments (tf) that are involved in the rewritten query and the slowdown (sd) compared to executing the query on the original database.
3. Instead of rewriting, the queries are cast to specific views set up in the trusted database server to recreate the original tables. For this method, the execution time (t) is measured, the number of involved table fragments (tf) and the slow down (sd) compared to the execution time of the same query on the original database. The number of involved table fragments is calculated by summing the number of table fragments that were necessary to create every view involved in the query.
4. If a query can be evaluated by a single database server (because it physically stores all the involved attributes), the query is directly cast to this server. For this method, only the execution time (t) is stated because it can be suspected, that their execution time is about the same as for the original database.

Figure 15 summarizes the results of the test runs while Table 9 shows the exact results. This evaluation shows the major advantage of the separation of duties approach. Because the columns of the tables are outsourced in plaintext, every query can potentially be executed. In particular, we are able to process all queries of the TPC-H benchmark. This is in contrast to approaches using property-preserving encryption: The MONOMI system [31] executes only 19 out of 22 TPC-H queries due to lacking support for views and text pattern matching; according to [31] the CryptDB system [27] executes only four out of the 22 queries.

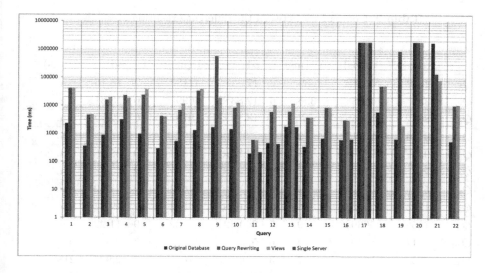

Fig. 15. Runtime results for TPC-H queries

However, we had to cancel two of the queries, namely Q_{17} and Q_{20}, because the timeout limit (30 min) was exceeded. Yet, the reason why these queries take so much time is not related to the vertically fragmented database as the timeout was also reached for the original non-fragmented database. Therefore, it can be concluded this issue is related to the PostgreSQL database engine which cannot find an adequate execution plan for those queries. Notably, [31] report the same problems when running the TCP-H queries: "Queries 17, 20, and 21 cause trouble for the Postgres optimizer: they involve correlated subqueries, which the optimizer is unable to handle efficiently".

As it was suspected, queries Q_{11}, Q_{12}, Q_{13} and Q_{16} that can be evaluated in a reasonable amount of time by a *single* server of the fragmented database can be executed in about the same time as in the non-fragmented database. For these 4 queries, a visibility constraint could be satisfied which perfectly illustrates the benefits of introducing those constraints. Interestingly, query rewriting and using views performed considerably worse for three of those 4 queries (Q_{11}, Q_{12}, Q_{13}). This is especially noticeable because rewriting the query also leads to a situation where the query involves only one database server but this is obviously not detected by PostgreSQL in conjunction with the foreign data wrapper extension postgres_fdw. This observation justifies a prior analysis of the queries as implemented in our distributed database client.

There is one query, Q_{21}, for which the execution time on the fragmented database is lower than the execution time for the non-fragmented database. For this query, the fragmented database probably profited from a better execution strategy that could be established by PostgreSQL due to the query rewriting or the use of the views. However, we assume that such situations occur very rarely in practice and are caused by PostgreSQL's execution strategy.

Table 9. Comparison between the different execution methods

Query	I	II				III			IV
	t (ms)	t (ms)	tr (ms)	tf	sd	t (ms)	tf	sd	t (ms)
Q_1	2267	40413	36	2	17.83×	41180	2	18.16×	n.a.
Q_2	353	4528	396	6	12.83×	4699	6	13.31×	n.a.
Q_3	861	15650	7	4	18.18×	19797	6	22.99×	n.a.
Q_4	3110	22917	12	4	7.37×	18571	4	5.97×	n.a.
Q_5	952	23658	5	7	24.85×	37036	10	38.9×	n.a.
Q_6	291	4167	1	2	14.32×	4039	2	13.88×	n.a.
Q_7	530	6817	14	6	12.86×	11653	9	21.99×	n.a.
Q_8	1305	33453	14	8	25.63×	38584	11	29.57×	n.a.
Q_9	1652	563036	9	7	340.82×	18532	9	11.22×	n.a.
Q_{10}	1417	8340	4	6	5.89×	12547	7	8.85×	n.a.
Q_{11}	193	595	7	3	3.08×	576	5	2.98×	218
Q_{12}	457	5921	4	2	12.96×	10399	4	22.75×	424
Q_{13}	1726	6103	2	2	3.54×	11821	4	6.85×	1696
Q_{14}	341	3721	2	3	10.91×	3765	3	11.04×	n.a.
Q_{15}	663	8368	3	2	12.62×	8685	4	13.1×	n.a.
Q_{16}	603	3054	5	3	5.06×	2983	6	4.95×	634
Q_{17}	Timeout	Timeout	57	2	n.a.	Timeout	3	n.a.	Timeout
Q_{18}	5998	50501	11	4	8.42×	51034	6	8.51×	n.a.
Q_{19}	646	859324	5	3	1330.22×	1927	3	2.98×	n.a.
Q_{20}	Timeout	Timeout	61	5	n.a.	Timeout	8	n.a.	n.a.
Q_{21}	1708506	136042	59	7	0.08×	79111	7	0.05×	n.a.
Q_{22}	534	10176	5	4	19.06×	10580	4	19.81×	n.a.

An interesting thing to notice is that query rewriting outperformed querying the views 13 times; querying the views was better for only 7 queries. Even more interesting, rewriting the queries performed better in 9 out of 12 times (ignoring the canceled queries) when the number of involved table fragments was lower than for the views. This illustrates the advantage of query rewriting over using views because unnecessary table fragments can be omitted with the former method. The overhead introduced by rewriting the queries is very small for all of the queries compared to the execution time and can therefore be neglected. Consequently, one can conclude that query rewriting is generally the better strategy than using views. However, if for some reason a rewritten takes very long to process, querying the views can potentially lower the execution time. An example for such a situation is query Q_{19}.

8   Conclusion and Future Work

In this article, we extended our separation of duties approach with which confidentiality in cloud databases can be enforced based on vertical fragmentation. Our approach enforces a security policy consisting of confidentiality constraints while at the same time respecting data dependencies, minimizing the amount of external cloud servers (the cardinality of the fragmentation) as well as maximizing the amount of satisfied visibility constraints (the constraints introduced to increase the utility of the resulting fragmentations). An implementation based on the provided theories was presented by translating the separation of duties problem into an integer linear program (ILP) representation and using an off-the-shelf solver to obtain a confidentiality-preserving fragmentation. In addition, we discussed our query rewriting approach, that enables an efficient distributed execution of queries on the fragments.

To show the feasibility of the separation of duties approach, based on the well-known TPC-E database schema the effects of different sizes of input data were evaluated. The evaluation of a TPC-H benchmark showed the major advantage of the separation of duties approach. As the columns of the database are stored in plaintext, it is possible to evaluate any database query, regardless of its complexity. Compared to encryption schemes, there is also no additional resource-intensive workload like decrypting the received data at the database user's site. Therefore, users of cloud databases who potentially run devices with a low computational power, especially benefit from this approach.

Several options for future work arise. Our approach is currently most applicable to situations where the constraint sets remain fixed over time. Studying certain classes of "allowed" modifications of these sets (confidentiality constraints, visibility constraints and dependencies) as well as their influences on security, data distribution and query execution is a major future topic which can be based on [3]. Moreover we plan to provide an in-depth analysis of different classes of integrity constraints similar to [2,4] as well as considering the query execution cost as an extra optimization goal. More generally in order to integrate our prior work on property-preserving encryption [34] we aim to analyze the combination of these encryption methods with separation of duties. Lastly it might be worthwhile to analyze the separation of duties approach in non-relational data models [36].

References

1. Aggarwal, G., et al.: Two can keep a secret: a distributed architecture for secure database services. In: The Second Biennial Conference on Innovative Data Systems Research (CIDR 2005) (2005)
2. Biskup, J., Preuß, M.: Database fragmentation with encryption: under which semantic constraints and a priori knowledge can two keep a secret? In: Wang, L., Shafiq, B. (eds.) DBSec 2013. LNCS, vol. 7964, pp. 17–32. Springer, Heidelberg (2013). https://doi.org/10.1007/978-3-642-39256-6_2

3. Biskup, J., Preuß, M.: Inference-proof data publishing by minimally weakening a database instance. In: Prakash, A., Shyamasundar, R. (eds.) ICISS 2014. LNCS, vol. 8880, pp. 30–49. Springer, Cham (2014). https://doi.org/10.1007/978-3-319-13841-1_3

4. Biskup, J., Preuß, M., Wiese, L.: On the inference-proofness of database fragmentation satisfying confidentiality constraints. In: Lai, X., Zhou, J., Li, H. (eds.) ISC 2011. LNCS, vol. 7001, pp. 246–261. Springer, Heidelberg (2011). https://doi.org/10.1007/978-3-642-24861-0_17

5. Bollwein, F.: CloudDBSOD Client. http://www.uni-goettingen.de/de/558180.html

6. Bollwein, F., Wiese, L.: Closeness constraints for separation of duties in cloud databases as an optimization problem. In: Calì, A., Wood, P., Martin, N., Poulovassilis, A. (eds.) BICOD 2017. LNCS, vol. 10365, pp. 133–145. Springer, Cham (2017). https://doi.org/10.1007/978-3-319-60795-5_14

7. Bollwein, F., Wiese, L.: Separation of duties for multiple relations in cloud databases as an optimization problem. In: Proceedings of the 21st International Database Engineering and Applications Symposium, pp. 98–107. ACM (2017)

8. Canim, M., Kantarcioglu, M., Inan, A.: Query optimization in encrypted relational databases by vertical schema partitioning. In: Jonker, W., Petković, M. (eds.) SDM 2009. LNCS, vol. 5776, pp. 1–16. Springer, Heidelberg (2009). https://doi.org/10.1007/978-3-642-04219-5_1

9. Chakravarthy, S., Muthuraj, J., Varadarajan, R., Navathe, S.B.: An objective function for vertically partitioning relations in distributed databases and its analysis. Distrib. Parallel Databases 2(2), 183–207 (1994)

10. Ciriani, V., De Capitani di Vimercati, S., Foresti, S., Jajodia, S., Paraboschi, S., Samarati, P.: Fragmentation and encryption to enforce privacy in data storage. In: Biskup, J., López, J. (eds.) ESORICS 2007. LNCS, vol. 4734, pp. 171–186. Springer, Heidelberg (2007). https://doi.org/10.1007/978-3-540-74835-9_12

11. Ciriani, V., De Capitani di Vimercati, S., Foresti, S., Jajodia, S., Paraboschi, S., Samarati, P.: Fragmentation design for efficient query execution over sensitive distributed databases. In: ICDCS, pp. 32–39. IEEE Computer Society (2009)

12. Ciriani, V., De Capitani di Vimercati, S., Foresti, S., Jajodia, S., Paraboschi, S., Samarati, P.: Keep a few: outsourcing data while maintaining confidentiality. In: Backes, M., Ning, P. (eds.) ESORICS 2009. LNCS, vol. 5789, pp. 440–455. Springer, Heidelberg (2009). https://doi.org/10.1007/978-3-642-04444-1_27

13. Ciriani, V., De Capitani Di Vimercati, S., Foresti, S., Jajodia, S., Paraboschi, S., Samarati, P.: Combining fragmentation and encryption to protect privacy in data storage. ACM Trans. Inf. Syst. Secur. (TISSEC) 13(3), 22 (2010)

14. Ciriani, V., De Capitani di Vimercati, S., Foresti, S., Jajodia, S., Paraboschi, S., Samarati, P.: Selective data outsourcing for enforcing privacy. J. Comput. Secur. 19(3), 531–566 (2011)

15. Ciriani, V., De Capitani di Vimercati, S., Foresti, S., Livraga, G., Samarati, P.: An OBDD approach to enforce confidentiality and visibility constraints in data publishing. J. Comput. Secur. 20(5), 463–508 (2012)

16. DBT-3. http://osdldbt.sourceforge.net/

17. De Capitani di Vimercati, S., Erbacher, R.F., Foresti, S., Jajodia, S., Livraga, G., Samarati, P.: Encryption and fragmentation for data confidentiality in the cloud. In: Aldini, A., Lopez, J., Martinelli, F. (eds.) FOSAD 2012-2013. LNCS, vol. 8604, pp. 212–243. Springer, Cham (2014). https://doi.org/10.1007/978-3-319-10082-1_8

18. De Capitani di Vimercati, S., Foresti, S., Jajodia, S., Livraga, G., Paraboschi, S., Samarati, P.: Fragmentation in presence of data dependencies. IEEE Trans. Dependable Secure Comput. **11**(6), 510–523 (2014)

19. De Capitani di Vimercati, S., Foresti, S., Jajodia, S., Paraboschi, S., Samarati, P.: Fragments and loose associations: respecting privacy in data publishing. Proc. VLDB Endow. **3**(1–2), 1370–1381 (2010)

20. Dwork, C.: Differential privacy: a survey of results. In: Agrawal, M., Du, D., Duan, Z., Li, A. (eds.) TAMC 2008. LNCS, vol. 4978, pp. 1–19. Springer, Heidelberg (2008). https://doi.org/10.1007/978-3-540-79228-4_1

21. Göge, C., Waage, T., Homann, D., Wiese, L.: Improving fuzzy searchable encryption with direct bigram embedding. In: Lopez, J., Fischer-Hübner, S., Lambrinoudakis, C. (eds.) TrustBus 2017. LNCS, vol. 10442, pp. 115–129. Springer, Cham (2017). https://doi.org/10.1007/978-3-319-64483-7_8

22. HammerDB. http://www.hammerdb.com/

23. Homann, D., Göge, C., Wiese, L.: Dynamic similarity search over encrypted data with low leakage. In: Livraga, G., Mitchell, C. (eds.) STM 2017. LNCS, vol. 10547, pp. 19–35. Springer, Cham (2017). https://doi.org/10.1007/978-3-319-68063-7_2

24. Hore, B., Jammalamadaka, R.C., Mehrotra, S.: Flexible anonymization for privacy preserving data publishing: a systematic search based approach. In: Seventh SIAM International Conference on Data Mining. SIAM (2007)

25. Jindal, A., Palatinus, E., Pavlov, V., Dittrich, J.: A comparison of knives for bread slicing. Proc. VLDB Endow. **6**(6), 361–372 (2013)

26. Özsu, M.T., Valduriez, P.: Principles of Distributed Database Systems. Springer, New York (2011). https://doi.org/10.1007/978-1-4419-8834-8

27. Popa, R.A., Redfield, C., Zeldovich, N., Balakrishnan, H.: CryptDB: processing queries on an encrypted database. Commun. ACM **55**(9), 103–111 (2012)

28. Sweeney, L.: k-anonymity: a model for protecting privacy. Int. J. Uncertain. Fuzziness Knowl.-Based Syst. **10**(05), 557–570 (2002)

29. Transaction Processing Performance Council: TPC-E Benchmark Version 1.14.0. http://www.tpc.org/tpce/

30. Transaction Processing Performance Council: TPC-H Benchmark Version 2.17.1. http://www.tpc.org/tpch/

31. Tu, S., Kaashoek, M.F., Madden, S., Zeldovich, N.: Processing analytical queries over encrypted data. In: Proceedings of the VLDB Endowment, vol. 6, pp. 289–300. VLDB Endowment (2013)

32. Waage, T., Homann, D., Wiese, L.: Practical application of order-preserving encryption in wide column stores. In: SECRYPT, pp. 352–359. SciTePress (2016)

33. Waage, T., Jhajj, R.S., Wiese, L.: Searchable encryption in Apache Cassandra. In: Garcia-Alfaro, J., Kranakis, E., Bonfante, G. (eds.) FPS 2015. LNCS, vol. 9482, pp. 286–293. Springer, Cham (2016). https://doi.org/10.1007/978-3-319-30303-1_19

34. Waage, T., Wiese, L.: Property preserving encryption in NoSQL wide column stores. In: Panetto, H., et al. (eds.) OTM 2017. LNCS, vol. 10574, pp. 3–21. Springer, Cham (2017). https://doi.org/10.1007/978-3-319-69459-7_1

35. Wiese, L.: Horizontal fragmentation for data outsourcing with formula-based confidentiality constraints. In: Echizen, I., Kunihiro, N., Sasaki, R. (eds.) IWSEC 2010. LNCS, vol. 6434, pp. 101–116. Springer, Heidelberg (2010). https://doi.org/10.1007/978-3-642-16825-3_8

36. Wiese, L.: Advanced Data Management for SQL, NoSQL, Cloud and Distributed Databases. DeGruyter/Oldenbourg, Munich (2015)

37. Xiao, Y., Xiong, L., Yuan, C.: Differentially private data release through multi-dimensional partitioning. In: Jonker, W., Petković, M. (eds.) SDM 2010. LNCS, vol. 6358, pp. 150–168. Springer, Heidelberg (2010). https://doi.org/10.1007/978-3-642-15546-8_11

38. Zakerzadeh, H., Aggarwal, C.C., Barker, K.: Managing dimensionality in data privacy anonymization. Knowl. Inf. Syst. **49**(1), 341–373 (2016)

39. Zhang, J., Xiao, X., Xie, X.: PrivTree: a differentially private algorithm for hierarchical decompositions. In: Proceedings of the 2016 International Conference on Management of Data, pp. 155–170. ACM (2016)

LPL, Towards a GDPR-Compliant Privacy Language: Formal Definition and Usage

Armin Gerl[1]([✉]), Nadia Bennani[2], Harald Kosch[1], and Lionel Brunie[2]

[1] DIMIS, University of Passau, Passau, Germany
{Armin.Gerl,Harald.Kosch}@uni-passau.de
[2] LIRIS, University of Lyon, Lyon, France
{Nadia.Bennani,Lionel.Brunie}@insa-lyon.fr

Abstract. The upcoming *General Data Protection Regulation (GDPR)* imposes several new legal requirements for privacy management in information systems. In this paper, we introduce LPL, an extensible Layered Privacy Language that allows to express and enforce these new privacy properties such as personal privacy, user consent, data provenance, and retention management. We present a formal description of LPL. Based on a set of usage examples, we present how LPL expresses and enforces the main features of the GDPR and application of state-of-the-art anonymization techniques.

Keywords: Anonymization · GDPR · LPL · Personal privacy
Privacy language · Privacy model · Privacy-preservation · Provenance

1 Introduction

Privacy is a research field which is tackled by different disciplines including computer and legal sciences. Each discipline has its own point of view on this complex topic. In computer science, privacy languages, in addition to express privacy rules, have been proposed to solve individual problem statements of privacy like informing users of the privacy settings of a website [1] or sharing and trading with (personal) data [2]. Furthermore, a privacy language is a data model of formal description which is machine-readable for automatic processing.

The *General Data Protection Regulation (GDPR)*, which will enter into force on 25th May 2018 [3, Art. 99 No. 2], is designed to standardise data privacy laws across Europe, to protect and empower all EU citizens (*Data Subjects*) data privacy and to rework the way organizations (*Controllers*) approach data privacy. Hereby, it advises to take a set of technical and organisational measures that could be summarized by two main principles, which are *Privacy by Design* and *Privacy by Default*, especially to protect *Data Subject Rights*.

We interpret *Privacy by Design*, which is an already existing concept that becomes now a legal requirement in the *GDPR*, as the requirement for a cross-domain definition of privacy policies which can be integrated in current business

© Springer-Verlag GmbH Germany, part of Springer Nature 2018
A. Hameurlain and R. Wagner (Eds.): TLDKS XXXVII, LNCS 10940, pp. 41–80, 2018.
https://doi.org/10.1007/978-3-662-57932-9_2

processes [4]. Therefore, privacy should be made available in all technical systems. To reach the *Privacy by Default* principal, it should be ensured that data access is permitted solely to persons and organizations that have the rights to access it or to which the *Data Subject* gives an explicit consent [3, Art. 25]. Additional legal aspects are listed in Sect. 2.1.

Our objective is to design a privacy language which aims to facilitate expressing legal requirements under the usage of privacy-preserving methodologies. With a formal definition of the privacy language we want to fulfill the principle of *Privacy by Design* by creating a machine-readable privacy policy for integration in technical systems. Furthermore, to fulfill the principle of *Privacy by Default*, we aim to cover all crucial privacy processes including the *Data Subject* giving its *consent* to the data processing, storage of personal data, transfer of personal data between *Controllers*, and privacy-preserving querying. Hereby, our proposed privacy language will serve as the base for a privacy-preserving framework supporting all mentioned processes. The privacy language presented in this paper allows expressing a static status of an organization, which we plan to extend by dynamic scenarios in future works.

To illustrate the purpose of our language, let's take the example of a *Data Subject* who registers an account for a service of *Controller C1*. Based on the consent, the personal data as well as our privacy language representing the legal privacy policy will be stored. According to the agreed privacy policy, the data will be transferred to *Controller C2* for statistical processing, whereas an additional privacy policy is created between both *Controllers* represented by our privacy language. The original privacy policy will then be appended to allow provenance for the personal data. *Controller C2* is processing personal data from several sources with different privacy policies as a service. Based on the requesting entity, *Controller C2* anonymizes the data according to the different individual privacy policies. Therefore, it can preserve privacy according to the legal regulations while delivering the best possible data utility. Additionally, both *Controllers* have to fulfill the *Data Subject Rights* given by the *GDPR*, e.g. disclosure of personal data. The legally required responses will be generated automatically based on our privacy language, reducing the workload for a *Controller*. To the best of our knowledge, there is no language that lets express and enforce the illustrated privacy-preserving features.

The main contribution of this paper is to present our *Layered Privacy Language (LPL)*. A formal description of its components is given. Then a set of usage patterns illustrating how policies are enforced are presented. The main focus is hereby on the *Query-based Anonymization*. Our goal with LPL is to model and enforce privacy policies, so that in Large-Scale Data and Knowledge-Centered-Systems it is possible to handle different personal privacy settings and therefore comply with the GDPR.

The remaining of the paper is structured as follows. In Sect. 2, considered aspects of privacy are listed and objectives for our proposed privacy language are derived from them. Section 3 reviews related works and positions LPL to them. Section 4 presents the formal description of LPL. Section 5 presents several usage patterns to illustrate several privacy aspects. Finally, Sect. 6 concludes and outlooks for future works.

2   Requirements

A privacy language should be able to express both legal and privacy-preserving requirements. Those requirements will be derived from the law and regulations and current state of the art of privacy-preservation methodology.

2.1   Legal View

We put the focus on the legal situation of Europe. A privacy policy or a privacy form can be translated as a set of rules describing how data has to be *processed*. Hereby, *processing* is broadly defined as collection, recording, organisation, structuring, storage, adaptation or alteration, retrieval, consultation, use, disclosure by transmission, dissemination or otherwise making available, alignment or combination, restriction, erasure or destruction of data [3, Art. 4 No. 2]. *Data Minimisation* denotes that only the minimum amount of data, which is necessary for the *processing* of the *purpose*, should be inquired from the *Data Subject* [3, Art. 5 No. 1 c)]. A privacy policy consists of several *purposes of the processing* [3, Art. 4 No. 9], describing what data is used, how it is used, when it will be deleted, who will use the data and if the data is anonymized [3, Art. 13]. Therefore, a privacy language has to be at least capable of modelling a set of purposes that have a set of data, set of data recipients, retention and the possibility to describe anonymized data [5]. Additionally several aspects of the European laws on privacy should be considered:

- *Consent:* A user has to give his *consent* for the *processing* of its data [3, Art. 6]. Hereby, the GDPR specifies that a *consent* has to be given freely, specific, informed and unambiguous [3, Art. 4 No. 11].
- *Personal Data:* The *GDPR* specifies *personal data* as any information that is related to an identified or identifiable natural person. This is a broad definition including among others name, location data, (online-)identifier and factors of a natural person [3, Art. 4 No. 1].
- *Purpose of the Processing:* The *processing* of *personal data* is only allowed for the defined purpose for which the user gave its *consent*. The *GDPR* specifies that personal data can only be collected and *processed* for legitimate *purpose of the processing* [3, Art. 5 No. 1 (b)]. The *purpose of the processing* is determined by the *Controller* which is a natural or legal person, public authority, agency or other body [3, Art. 4 No. 7]. Exceptions to this are also possible but will not be further discussed [3, Art. 6].
- *Retention:* According to the *GDPR*, personal data has to be deleted when it is no longer necessary for the *purpose of the processing* for which it was collected, which is a part of the *'right to erasure'* or *'right to be forgotten'* of the data subject [3, Art. 14 No. 1]. Therefore, deletion of personal data is

strictly bound to the *purpose of the processing*. The policy is to delete data when it is no longer necessary for the *purpose of the processing* or the *purpose of the processing* is completed. For example if the *purpose of the processing* is solely to use the e-mail for the newsletter, then data is revoked, once the newsletter subscription is completed.

– *Data Subject Rights:* The *GDPR* defines several *Data Subject Rights* including among others the *'right of access by the data subject'* [3, Art. 15], *'right to rectification'* [3, Art. 16], *'right to erasure'* [3, Art. 17], and *'right to object'* [3, Art. 21] giving the *Data Subject* several rights that have to be considered. For example, if the *Data Subject* has given its *consent* for the *processing* of the *personal data* to a *Controller*, then the *Data Subject* has also the right to demand the deletion of the *personal data* if there is a valid reason for it [3, Art. 17].

This breakdown of the legal regulations omits further exceptions and special cases for each of the mentioned points in favour of the scope of this paper. We are further aware that the European law constraints are not compliant with other regulations like the Health Insurance Portability and Accountability Act (HIPAA), but similar basic concepts can also be found in these regulations [6].

2.2 Privacy-Preserving View

We will focus on *Anonymization and Privacy Model Requirements, Data Storage Requirements* and *Personal Privacy Requirements* to describe the considered requirements for our privacy language. We are aware that further requirements from other privacy research fields like database trackers [7] could be added, but those would be more relevant for a privacy framework, than a privacy language, and therefore are out of scope for this paper.

Anonymization and Privacy Model Requirements. There are several privacy models like *k-Anonymity* [8], *l-Diversity* [9] or *t-Closeness* [10] defining the properties a data-set must have to prevent re-identification. To explain the requirement regarding the privacy-preservation item, let's take in this section the example of the *k-Anonymity* model. The properties of the privacy models are usually adjusted by one or several parameters. Illustrated on *k-Anonymity*, the parameter k defines for a data-set, that for each QID-group, at least k records have to exist [8]. A QID (quasi-identifier) is hereby an attribute which can, in combination with other QID attributes, be used for identification, but can by itself not used for identification.

Based upon the chosen value, the properties of the anonymized data-set in terms of utility and privacy are influenced.

Utility describes the data quality of an anonymized data-set in relation to the original data-set. The quality of the data can hereby be highly dependent on the context in which the data-set is supposed to be used. But in general *utility* can be measured by *Accuracy, Completeness* and *Consistency. Accuracy* measures the similarity, e.g. loss of information, between the anonymized value and the original value. *Completeness* measures the missed data in the anonymized dataset. *Consistency* measures if the relationships between data items is preserved. Based upon those measurements several methods have been developed for *utility*, e.g. for *k-Anonymity* the *height* metric is used [11].

A trade-off between privacy and utility has to be found which is defined by the privacy model as well as the corresponding parameters. An open question is who decides on the privacy model and its parameters. The choice of the privacy-preserving parameters defines the privacy of a data-set. Should those settings exclusively be decided by a privacy expert, e.g. a privacy officer in a company, a national authority or can this even be influenced or set by a user, is the question.

A higher value for parameter k results in higher privacy for the data-set. With increasing privacy of the data-set, the more likely it is that anonymization has to be applied on the data which has a negative impact on the *utility*. It has to be considered that an overestimated value will result in an undesired loss of *utility*. An underestimated value will result in insufficient privacy. This leads to the requirement that the definition of diverse privacy models including their privacy-preserving parameters has to be supported. This includes that data attributes have to be able to be assigned to a specified privacy group (e.g. Quasi-identifier) to enable a correct application of the privacy model. Because it is an open question which entity (privacy officer, *Data Subject*, or both) should influence the definition of the privacy-preserving parameters, we consider that those parameters can be influenced by both.

Data Storage and Transfer Requirements. Thus far we considered privacy only for a homogeneous data-set, but privacy has also to be considered in data-warehouses, and other storage solutions, which implicates different data-sources and queries. Each query-result can be imagined as a data-set for which privacy has to be considered. Therefore, the data can be anonymized at different points in time of privacy-preserving data-warehousing. For example it is possible that the source data is already anonymized before it is integrated in the data-warehouse. Alternatively, it is also possible to anonymize the data for each query conducted on the data-warehouse. In general, the possibilities for the point of anonymization in a data-warehouse scenario are *anonymized sources*, *pre-materialization anonymization*, *post-materialization anonymization* and *query-based anonymization*. Each of the approaches has its own advantages and disadvantages. It is shown that *post-materialization anonymization* has significant advantages over *anonymized sources* and *pre-materialization anonymization* in terms of data quality. If an untrusted data publisher model is selected then anonymized data sources are a necessity and therefore the *post-materialization anonymization* approach cannot be chosen. Experiments for *query-based anonymization* have not been conducted and therefore cannot be compared [12]. Based upon these results, we assume that a *as-late-as-possible anonymization* is advantageous, which we consider as a requirement.

In (privacy-preserving) data warehousing the data is combined in a single warehouse system from one or more data-sources and queried by data-recipients.

Generally speaking data will be *transferred, materialized, anonymized* and *queried*. This process may be run through several times, thus the origin of the data may be lost if it is not explicitly tracked. Therefore, it is required to store for each data record the corresponding privacy policy. Assuming that different data-sources have inherent different privacy policies, this process will lead usually to a data warehouse with diverse privacy policies that have to be considered. For example, we assume data-sources *source1* and *source2*, whereas *source1* delivers data under privacy policy *policy1* and *source2* under *policy2*. Data from both data-sources is combined in data-warehouse *warehouse1* including their corresponding privacy policies *policy1* and *policy2*. Therefore, it is a requirement that privacy policies, related to a specific data record, can be stored and transferred with the data. Hereby, the data should be stored as long as possible in its raw form to support a *as-late-as-possible anonymization*.

Furthermore, we are aware that sequential queries of a database have to be considered for privacy preservation. Each release can hereby contain a different set of attributes. The combination of those attributes, which are retrieved over time, may allow the identification of a *Data Subject* and therefore cause a privacy issue [7]. Although we are aware of this issue, we will not consider it for *Query-based Anonymization* within the scope of this paper.

Personal Privacy Requirements. The approach of allowing a user to set his *Personal Privacy Preferences* has been addressed in [13]. This approach gives the user the control over its privacy settings. To be more specific, it considers the minimum necessary anonymization of the data and therefore retains the maximum *utility* of the data. This approach considers *Personal Privacy Preferences* in the anonymization process of the data [13]. But we also consider the privacy model as part of the personal privacy settings. Therefore, an approach to find the minimum necessary privacy model and value can be derived for a data-set, which we denote as a requirement.

Additionally, we consider that it is possible that records from a data-source with personal privacy policies exist [13]. Therefore, the diversity of privacy policies that has to be considered rises. When the data is queried and transferred to a data recipient, it is possible that new privacy policies, representing additional privacy policies, of the queried data-warehouse are applied to the data-set and will be mixed up with the previous privacy policies, which can cause conflicts. This is not only restricted to a data-warehouse scenario, like mentioned before, but for every transfer of personal data. With every transfer of data it is possible, if not prevented, that the original *Data Subject* can no longer be identified explicitly, but its personal data is still processed. Therefore, a loss in provenance occurred. If the *Data Subject* wants to exercise his *Data Subject Rights* or the *Controller* has to prove the origin of the processed data it will no longer be possible. This has to be prevented. Consequently, we denote *Provenance* as a requirement.

This leads to the requirement of defining personal privacy within a privacy language on such a fine-grained level that each attribute may be influenced by both the privacy officer defining the maximum allowed anonymization and the *Data Subject* defining its own minimal privacy settings. Considering that several personalized privacy policies may be transferred and aggregated it is necessary to be able to track the origin of the data and therefore provenance has to be implemented within a privacy language.

2.3 Objectives

Based upon the legal and privacy-preserving view, we formulated the following objectives for a privacy language both representing legal privacy policies and ensuring privacy utilizing privacy-preserving methods.

It should be able to layer privacy policies to track the origin of the data and therefore enable *Provenance* for the personal data. Hereby, data-source and data-recipient should be differentiated to be able to grant fine-grained data processing rules. This has also to include the processing of *Data Subject Rights*, which are given by the law. The structure of the privacy language should match the structure of a privacy policy which is based on purposes, describing the circumstances of the data usage. Therefore, for each purpose it should be possible to describe the data that is processed, the data-recipient and retention. It is required that privacy for a data-set can be specified utilizing privacy models. But also privacy settings for single data fields are required to enable fine-grained personal privacy. Therefore, both a minimum level, defining privacy settings of the *Data Subject*, and maximum level, defining the upper limit for the *Controller*, are required. Finally, the privacy language should support the user consent on data access in a legal and human-readable way, whereas it has to considered that multiple languages are supported. Summarizing a privacy language should fulfill the following requirements:

- Differentiation between data-source and data-recipient to enable fine-grained access-control
- Modelling of purpose-based privacy policies with modelling of: data, retention and anonymization enabling personal privacy and privacy models
- Layering of privacy policies to ensure provenance
- Human-readability

In the following we will compare related works according to our requirements.

3 Related Works

We define a classification for privacy languages based on a broad literature research as well as on our previously defined requirements, which we then apply on a set of privacy languages to demonstrate a research gap.

Several privacy languages have been proposed in the literature, each with their own distinct purposes. Although they are classified as privacy languages by other works [14, 15], we do not see a strong focus on *Privacy* (in a legal sense)

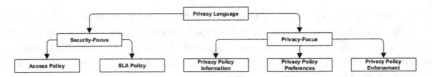

Fig. 1. Categorizes for classification of privacy languages.

in every of them. Therefore, we developed a classification of privacy languages according to their intended purposes (see Fig. 1). Hereby, we use the *Privacy* for purposes in which the languages deal with legal aspects of privacy. We differentiate between five intended purposes, that we could identify by a broad literature research.

For *Access Policy*, policies for access control are implemented, such as *XACL* [16], *Ponder* [17], *Rei* [18], *Polymer* [19], *SecPAL* [20], *AIR* [21], *XACML* [22] and *ConSpec* [23]. For *Service Level Agreement (SLA) Policy*, agreements or contracts for B2B processes are implemented, such as *SLAng* [24,25] and *USDL* [26]. For *Privacy Policy Information*, policies are implemented to (only) inform about their contents, such as *P3P* [27] and *CPExchange* [28]. For *Privacy Policy Preferences*, personal privacy settings or preferences (of e.g. users) are modelled to be matched against policies, such as *APPEL* [29] and *XPref* [30]. For *Privacy Policy Enforcement*, policies are modelled and implemented to enforce privacy policies, such as *DORIS* [31], *E-P3P* [32], *EPAL* [33], *PPL* [34], *Jeeves* [35], *Geo-Priv* [36], *Blowfish Privacy* [37], *Appel* [38], *P2U* [2] and *A-PPL* [39].

Additionally, we analyse if the privacy languages consider several topics, that we derived from our requirements, which will be detailed as follows. For *Purpose-oriented*, the *purpose* of the *processing* of data is modelled as a high-level process and not only as low-level *CRUD* operations. For *Data-oriented*, each *data* can modelled uniquely and not only as (pre-defined) groups of data. For *Retention*, rules for automatic deletion of data or data-sets based on the retention have to be modelled. The possibility of an active deletion request, issued by the *Data Subject*, does not fulfill this criteria. For *Access-Control*, mechanisms for authentication and authorization have to be enabled by the model. For *Human-Readability*, the model should allow a human-readable presentation, so that the *Data Subject* is informed about the content. For *Privacy Model*, the minimal privacy properties of the data-set for a specific *purpose* have to be modelled. For *Personal Privacy*, the *Data Subject* should be able to dissent the use of data for a specific *purposes* or the *processing* of a specific *purpose*. Furthermore the anonymization of data for a specific *purpose* should be able to be influenced. For *Provenance*, after data has been transferred between (multiple) *Controllers* the original *Data Subject* should still be identifiable, so that this *Data Subject* can enforce his *Data Subject Rights*.

Based on the presented classification, we analysed a broad range of privacy languages. An broad and comprehensive overview is shown in Table 1. It can be observed that most privacy languages, which categorized as *Privacy Policy Enforcement*, are *Purpose-oriented* and *Data-oriented*. Furthermore, most

Table 1. Overview over fulfilled objectives for a privacy language combining both the legal and privacy-preserving views on privacy.

Category	Privacy language	Purpose-oriented	Data-oriented	Retention	Access-control	Human-readability	Privacy model	Personal privacy	Provenance
Access Policy	XACL	x	x		x				
	Ponder	x			x				
	Rei	x			x				
	Polymer	x							
	SecPAL		x		x				
	AIR	x	x		x	x			
	XACML	x	x		x				
	ConSpec	x			x				
SLA Policy	SLAng	x	x	x					
	USDL		x			x			
Privacy Policy Information	P3P	x	x	x	x				
	CPExchange	x	x	x	x				
Privacy Policy Preferences	APPEL	x	x						
	XPref	x	x						
Privacy Policy Enforcement	DORIS	x	x		x				
	E-P3P	x	x	x	x				
	EPAL	x	x		x				
	PPL	x	x	x	x				
	Jeeves	x	x		x				
	Geo-Priv	x	x	x	x		x		
	Blowfish Privacy	x	x				x		
	Appel	x			x				
	P2U	x	x	x	x				
	A-PPL	x	x	x	x				

consider the topics *Retention* and *Access-Control*. Both *Geo-Priv* and *Blow-fish Privacy* specify the anonymization of personal data and therefore deal with the topic *Privacy Model*. The topics *Human-Readability*, *Personal Privacy* and *Provenance* are not dealt with by any of the privacy languages categorized as *Privacy Policy Enforcement*. The topic *Human-Readability* is only dealt with by *AIR* and *USDL*.

In summary, there is a lack of the legal and privacy-preserving requirements in the design of privacy languages, that we will consider. It is also mentionable that the representation of legal privacy policies (especially according to the *GDPR*) has not been the intention of any of the described privacy languages and has also not been done according to our knowledge. In the following, we give a formal definition for LPL implementing the described requirements.

4 Layered Privacy Language (LPL)

In this section, we present a formal description of our *Layered Privacy Language (LPL)* which satisfies the requirements presented in Sect. 2. The structure and the components of the language are depicted in Fig. 2, whereas attributes are omitted. All the elements presented in the diagram are described, including their attributes, in the following subsections. For clarity of the description, Table 2 gives for each element, notations that will be used for a single element, a subset of elements and the complete set.

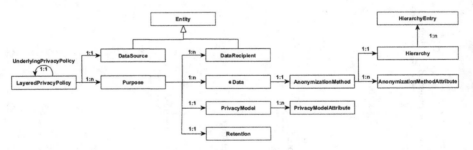

Fig. 2. Overview of the structure of LPL. Attributes are omitted for better readability.

4.1 Layered Privacy Policy

The root-element of our privacy language is the *LayeredPrivacyPolicy*-element *lpp*, which represents a privacy policy (legal contract), e.g. between a user and a company. Only a single *lpp* is supposed to be defined for a LPL compliant file, e.g. privacy policy. A *LayeredPrivacyPolicy*-element

$$lpp = (version, name, lang, ppURI, upp, ds, \widehat{P}) \tag{1}$$

is a tuple consisting of the following attributes:

- *version:* Version number for future version management of LPL.
- *name:* A textual representation of the privacy policy name.
- *lang:* Defining the language of the human-readable description in the LPL privacy policy.
- *ppURI:* A link to the legal privacy policy to assure compliance with the current law, which is implemented as a static human-readable description of the privacy policy.

Additionally, each *LayeredPrivacyPolicy lpp* can have a reference to an *UnderlyingPrivacyPolicy upp*. Let an *UnderlyingPrivacyPolicy*-element be

$$upp = (version, name, lang, ppURI, upp', ds, \widehat{P}) \qquad (2)$$

where *upp'* is another *UnderlyingPrivacyPolicy*-element denoting a previously consented privacy policy. The set of all *upp* elements is denoted by UPP and $\widehat{UPP}$ denotes a subset of UPP.

Table 2. Overview over all elements, their formal definition and a reference to their definition. Bold styled sets are tuples, which inherit an order.

Element	Single element	Subset of elements	Set of elements	Definition reference
LayeredPrivacyPolicy	lpp	$\widehat{LPP}$	LPP	Section 4.1
UnderlyingPrivacyPolicy	upp	$\widehat{UPP}$	UPP	Section 4.1
Purpose	p	$\widehat{P}$	P	Section 4.2
Entity	e	$\widehat{E}$	E	Section 4.3
DataSource	ds	$\widehat{DS}$	DS	Section 4.3
DataRecipient	dr	$\widehat{DR}$	DR	Section 4.3
Retention	r	$\widehat{R}$	R	Section 4.4
PrivacyModel	pm	$\widehat{PM}$	PM	Section 4.5
PrivacyModelAttribute	pma	$\widehat{PMA}$	PMA	Section 4.5
Data	d	$\widehat{D}$	D	Section 4.6
AnonymizationMethod	am	$\widehat{AM}$	AM	Section 4.7
AnonymizationMethodAttribute	ama	$\widehat{AMA}$	AMA	Section 4.7
Hierarchy	h	$\widehat{H}$	H	Section 4.7
HierarchyEntry	he	$\widehat{\mathbf{HE}}$	$\mathbf{HE}$	Section 4.7

This allows to create layers of privacy policies to satisfy the objective of being able to track privacy policies over multiple entities.

Let the ('most underlying') *leaf-LayeredPrivacyPolicy*

$$lpp_{leaf} = (version, name, lang, ppURI, \emptyset, ds, \widehat{P}) \qquad (3)$$

be the first privacy policy for which a consent is given for. In other words, the *LayeredPrivacyPolicy* with no *UnderlyingPrivacyPolicy* is the initial privacy policy, which is usually a consent between an user and a legal entity. If an additional privacy policy lpp_{new}, e.g. for a data-transfer to a third party, has to be added to an existing $lpp_{existing}$, then the $lpp_{existing}$ will be wrapped by lpp_{new}. This results in

$$lpp_{existing} = (version, name, lang, ppURI, \emptyset, ds, \widehat{P}) \tag{4}$$

$$lpp_{new} = (version, name, lang, ppURI, lpp_{existing}, ds', \widehat{P'}) \tag{5}$$

which is valid for each additional added privacy policy lpp'_{new}. Hereby, the data source (ds') will be the data recipient (dr) of $lpp_{existing}$ and $\widehat{P'}$ can be $\widehat{P}$ or a subset of it. The *DataSource*-element ds and a set $\widehat{P}$ of *Purpose*-elements will be described in the following. The set of all lpp elements is denoted by LPP and $\widehat{LPP}$ denotes a subset of LPP.

4.2 Purpose

The *Purpose*-element p, representing a legal *purpose of the processing*,

$$p = (name, optOut, required, descr, \widehat{DR}, r, pm, \widehat{D}) \tag{6}$$

is a tuple consisting of the following attributes:

- *name:* A textual representation of the identifying name, e.g. 'marketing'. In the set of purposes there should be no duplicate names.
- *optOut:* A boolean defining if the *Purpose* is opt-out for *true* or opt-in for *false*. Opt-out implies that the user has to actively deny this purpose. In the opposite, opt-in implies that the user has to actively accept this purpose.
- *required:* A boolean defining if the *Purpose* has to be accepted by the user. If the user does not accept a required *Purpose* then there cannot be a consent for the corresponding lpp.
- *descr:* A human-readable textual representation of the purpose expressed in the language defined by the language *lang* of lpp.

Moreover, each *Purpose* is linked to a set $\widehat{DR}$ of dr, one *Retention*-element r, optionally one *PrivacyModel*-element pm and a set $\widehat{D}$ of d. The set of all p elements is denoted by P and $\widehat{P}$ denotes a subset of P. It is important to note that the set $\widehat{P}$ may be empty or consist of contradictory purposes, which is valid for the structure but illogical for a privacy policy.

4.3 Entity

The *Entity*-element e, representing persons, companies or any other entity that has processing-right on the data,

$$e = (name, classification, authInfo, type) \tag{7}$$

is a tuple consisting of the following attributes:

- *name:* Used for authorization in access control.
- *classification:* Classifies the *Entity* in either *Person* or *Legal Entity.*
- *authInfo:* Used for authentication of the *Entity*, e.g. a hashed password.
- *type:* Either *DataSource* or *DataRecipient.*

The set of all *e* elements is denoted by E and $\widehat{E}$ denotes a subset of E. The *Entity*-element inherits the following 2 elements, which do not add additional attributes, but are used for better readability of LPL.

DataSource. The *DataSource*-element *ds* inherits from *Entity*, whereas the *type* is set to the corresponding value.

$$ds = (name, classification, authInfo, \text{'DataSource'}) \qquad (8)$$

The *DataSource*-element describes the current authority granting *DataRecipients* the *processing* of data, based upon its own processing-rights. For example this can be the user (person) for whom the personal data is dedicated to or a company (legal entity) that has collected the personal data for a specific purpose. The set of all *ds* elements is denoted by DS and $\widehat{DS}$ denotes a subset of DS.

DataRecipient. The *DataRecipient*-element *dr* inherits from *Entity*, whereas the *type* will be set to *DataRecipient*.

$$dr = (name, classification, authInfo, \text{'DataRecipient'}) \qquad (9)$$

The *DataRecipient*-element represents the authority that gets specific processing-rights (defined by the *Purpose*) granted. This can be a person or a legal entity. For example given the *DataSource*-element representing the user (person) which the personal data is referring to, then this authority can grant the *DataRecipient* all processing-rights via $\widehat{P}$. Assuming ds_C represents a *Controller* C that has collected the data from a user ds_U under specific processing-rights $\widehat{P}_C$ and wants to grant a third party dr_T processing-rights $\widehat{P}_T$, then ds_C can only grant dr_T the usage within the limits of its own processing-rights $\widehat{P}_T \subseteq \widehat{P}_C$. It has to be noted that the processing-rights of ds_C are a subset of the processing-rights of the user, who has all the processing-rights $\widehat{P}_T \subseteq \widehat{P}_C \subseteq \widehat{P}_U$ The set of all *dr* elements is denoted by DR and $\widehat{DR}$ denotes a subset of DR.

4.4   Retention

The *Retention*-element *r* defines when the described data has to be deleted.

$$r = (type, pointInTime) \qquad (10)$$

The element consists of the following attributes:

- *type:* Describing the general condition of the retention. Possible values are *Indefinite*, *AfterPurpose* and *FixedDate*.

– *pointInTime:* Textual representation describing the exact conditions for the retention.

Depending on the *type*, the *pointInTime* has diverse meanings. The *type Indefinite* without a value for *pointInTime* defines that there is no time constrained for the deletion of the data. The *type AfterPurpose* defines that after the completion of the corresponding purpose p the data has to be deleted within the time-frame specified by *pointInTime*. Lastly the *type FixedDate* in combination with *pointInTime* explicitly defines the date for the deletion of the data within the corresponding p. The set of all r elements is denoted by R and $\widehat{R}$ denotes a subset of R.

4.5   Privacy Model

The *PrivacyModel*-element pm specifies the privacy conditions that have to be fulfilled for the data in a data-set. This element can be given but it is not mandatory. Alternatively, privacy can also be defined by *AnonymizationMethod*-element, defining personal privacy, or even omitted if not necessary, e.g. when the p does not describe any personal data.

$$pm = (name, \widehat{PMA}) \tag{11}$$

The applied privacy model is defined by the *name*, e.g. *k-Anonymity* [8] or *l-Diversity* [9]. Each privacy model can have a set of *PrivacyModelAttribute*-elements $\widehat{PMA}$. Currently, we limit to one privacy model pm for each purpose p. It may be a requirement that more than one privacy model is applied to a data-set [40], which would be a possible future extension. The set of all pm elements is denoted by PM and $\widehat{PM}$ denotes a subset of PM.

Privacy Model Attribute. A *PrivacyModelAttribute*-element pma, represents the configuration of a privacy model,

$$pma = (key, value) \tag{12}$$

is a tuple of the following attributes:

– *key:* Definition of a variable that is required by the correlating pm, e.g. k for *k-Anonymity*.
– *value:* Definition of the actual variable content, e.g. for k the value '2', which describes that there have to be at least two records within the same *QID*-set values to preserve the required k-anonymity property [8]

The set of all pma elements is denoted by PMA and $\widehat{PMA}$ denotes a subset of PMA. The decision for utilizing $\widehat{PMA}$ can be explained by the existence of privacy model (e.g. i X, Y-*Privacy* [41]) that support more than one variable.

4.6   Data

The *Data*-element d, representing a data field that is concerned by a purpose p,

$$d = (name, dGroup, dType, required, descr, pGroup, am) \qquad (13)$$

is a tuple of the following attributes:

- *name:* Distinct name for the stored data field. A duplicate *name* within a $\widehat{D}$ of a *Purpose* is not allowed. Because this could lead to discrepancies in the processing, like it is possible with P3P *DATA-Elements*. P3P allows to define contrary rules for the same data element within one purpose. This made the determination of the valid rule unfeasible [42].
- *dGroup:* A textual representation of a logical data group. No predefined values are given. The logical data group can be used to specify data in e.g. a procedure directory, which usually does not refer to each data field but groups of it [3, Art. 30]. This enables to validate procedures directories [3] with a privacy policy automatically or even create one beforehand. E.g. data elements representing title, prename and surname of a person could have the *dGroup* ‘name’.
- *dType:* Defines the type of data that the attribute has. Possible types are *Text, Number, Date, Boolean, Value Set* for a set of predefined values and *Other* for any data type that doesn't fit the aforementioned types.
- *required:* A boolean defining if the data d to be accepted or could be neglected by the user. If the user does not accept a required d then the corresponding p will not be accepted. If the p is required, then the whole privacy policy *lpp* is not accepted.
- *descr:* A human-readable description of the data field. Possible notes on the anonymization, that is applied, can be added for better understanding. For example, the *age* will be only analysed in ranges from ‘0–50’ and ‘50–100’.
- *pGroup:* This is the classification of the data field in *Explicit, QID, Sensitive* and *Non-Sensitive*. The processing of the data field by the privacy models is based upon this classification. E.g. for *k-Anonymity* the value of a data field which is classified *Explicit*, has to be deleted [8].

The *AnonymizationMethod*-element am defines the minimum anonymization for the data enabling personal privacy. The set of all d elements is denoted by D and $\widehat{D}$ denotes a subset of D.

4.7   Anonymization Method

The *AnonymizationMethod*-element am, represents the anonymization that is applied on a data,

$$am = (name, \widehat{AMA}, h) \qquad (14)$$

is a tuple of the following attributes. The *name* represents the chosen anonymization method. There are several methods available, for example *Deletion, Suppression* or *Generalization*. Additionally each *AnonymizationMethod* has a set of *AnonymizationMethodAttributes*-elements $\widehat{AMA}$ and optionally a *Hierarchy*-element h. The set of all am elements is denoted by AM and $\widehat{AM}$ denotes a subset of AM.

Anonymization Method Attribute. An *AnonymizationMethodAttribute*-element ama, represents the configuration of a anonymization method,

$$ama = (key, value) \tag{15}$$

is a tuple of the following attributes:

- *key:* Definition of a variable that is required by the correlating am. Additionally, defining a maximum and minimum *Anonymization Level* is possible.
- *value:* Definition of the actual variable content.

The *key 'Minimum Level'* defines the minimum *Anonymization Level* that is applied to the actual data when used for a specific purpose. This allows to anonymize *as-late-as-possible*. The *key* with the value *'Maximum Level'* defines the maximum level of anonymization that is applied and therefore gives the creator of the privacy policy the possibility to define requirements for the anonymization. The set of all ama elements is denoted by AMA and $\widehat{AMA}$ denotes a subset of AMA.

Hierarchy. The *Hierarchy*-element h, saving all possible pre-calculated values for one data field and the correlating anonymization method. The hierarchy h will be used during the anonymization both for the *Minimum Anonymization*, enabling personal privacy, and the *Application of the Privacy Model*.

$$h = (\widehat{\mathbf{HE}}) \tag{16}$$

Each h consists of a tuple of *HierarchyEntry*-elements $\widehat{\mathbf{HE}}$, representing each entry in the hierarchy. We denote $h.length$ as the amount of he elements in $\widehat{\mathbf{HE}}$.

$$h.length = |\widehat{\mathbf{HE}}| \tag{17}$$

The tuple of all he elements is denoted by $\mathbf{HE}$ and $\widehat{\mathbf{HE}}$ denotes a sub-tuple of $\mathbf{HE}$. We decided for the calculation and storage of the possible anonymized values within the privacy language over the calculation of the anonymized value per query. The hierarchy is optional as not every *AnonymizationMethod*, e.g. *Deletion*, will have several possible values. Additionally it may also not be suitable for all use cases to store pre-calculated values. The set of all h elements is denoted by H and $\widehat{H}$ denotes a subset of H. Next section presents several usage patterns of LPL privacy policies.

5   Usage of LPL

In this section we present the life-cycle of LPL. Furthermore, we present *Query-based Anonymization*, *Provenance* and *Retention* based on LPL.

For the sake of clarity and to avoid redundancy, for each usage pattern we will define progressively the elements examples of LPL that are mandatory for the illustration. Each previously defined element could be referenced in further usage patterns if required. Only the relevant attributes will be instantiated for better readability.

5.1   Life-Cycle

For LPL, we present the life-cycle steps through the following scenario: A company e_{C1} wants to create a new web-service which collects and uses personal information. Therefore, e_{C1} creates a legal privacy policy that a user e_{U1}, using the service, has to accept. In this case ds_{U1} is the *'DataSource'* and dr_{C1} is the *'DataRecipient'*. Optionally, the data that is collected by e_{C1} could be transferred to a third party e_{C2} for a specific usage. Therefore, a contract between e_{C1} and e_{C2} has to be concluded for the data transmission, whereas ds_{C1} is the *'DataSource'* and dr_{C2} is the *'DataRecipient'*.

$$ds_{U1} = (\text{'U1'}, \text{'Person'}, publicKey_{U1}, \text{'DataSource'}) \tag{18}$$

$$dr_{C1} = (\text{'C1'}, \text{'Legal Entity'}, publicKey_{C1}, \text{'DataRecipient'}) \tag{19}$$

$$ds_{C1} = (\text{'C1'}, \text{'Legal Entity'}, publicKey_{C1}, \text{'DataSource'}) \tag{20}$$

$$dr_{C2} = (\text{'C2'}, \text{'Legal Entity'}, publicKey_{C2}, \text{'DataRecipient'}) \tag{21}$$

The usage of LPL in this scenario can be separated into the following steps (see Fig. 3):

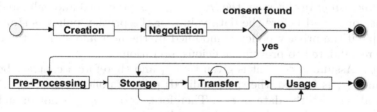

Fig. 3. Life-cycle of LPL.

1. *Creation:* Company e_{C1} converts the legal privacy policy to an LPL privacy policy lpp_{raw}. Hereby e_{C1} defines which *Purpose-* and *Data-* elements are necessary for the usage of the web-services as well as all other elements. In this case, a privacy policy will be transferred describing personal data to

be used by dr_{C1} and dr_{C2} for the purpose '*Marketing*'. This includes the anonymization for '*postal-code*', but not for '*salary*'.

$$lpp_{raw} = (version, \text{'LPP1'}, lang, ppURI, \emptyset, \emptyset, \{p_{U1}\}) \tag{22}$$

$$p_{U1} = (\text{'Marketing'}, optOut, required, descr, \{dr_{C1}, dr_{C2}\}, r, pm, \widehat{D}_1) \tag{23}$$

$$\widehat{D}_1 = \{d_{postal}, d_{salary}\} \tag{24}$$

$$d_{postal} = (\text{'postal-code'}, dGroup, dType, required, descr, \text{'QID'}, am_1) \tag{25}$$

$$am_1 = (\text{'Suppression'}, \{ama_1, ama_2, ama_3, ama_4\}, \emptyset) \tag{26}$$

$$ama_1 = (\text{'Suppression Replacement'}, \text{'*'}) \tag{27}$$

$$ama_2 = (\text{'Suppression Direction'}, \text{'backward'}) \tag{28}$$

$$ama_3 = (\text{'Minimum Level'}, \text{'2'}), ama_4 = (\text{'Maximum Level'}, \text{'4'}) \tag{29}$$

$$d_{salary} = (\text{'salary'}, dGroup, dType, required, descr, \text{'Sensitive'}, \emptyset) \tag{30}$$

2. *Negotiation:* The privacy policy lpp_{raw} is presented to the user e_{U1} via a user-interface, enabling an informed and voluntary consent. The user-interface should also give e_{U1} the possibility to dissent with defined parts of lpp_{raw} and still be able to form a contract with e_{C1}, whereas a personalized privacy policy $lpp_{ds_{U1}\text{-}dr_{C1}}$ is created. This leads to the insertion of ds_{U1}. If no consent is found the user cannot use the web-service and no data nor privacy policy of e_{U1} will be stored.

$$lpp_{ds_{U1}\text{-}dr_{C1}} = (version, \text{'LPP1'}, lang, ppURI, \emptyset, ds_{U1}, \{p_{U1}\}) \tag{31}$$

A user-interface that allows to personalize the LPL privacy policy has been developed and evaluated. It focuses on the consent or dissent to purposes. Further features, like the personalization of minimum anonymization for specific data or the consent or dissent to specific data, are planned for future work.

3. *Pre-Processing:* In this step, the $lpp_{ds_{U1}\text{-}dr_{C1}}$ is processed and validated. This step is conducted before the data values or the privacy policy is stored. For example if the privacy policy is modified by the user and stored, then it has to be re-validated to prevent malicious alterations.

4. *Storage:* Assuming consent is given by e_{U1} and therefore a contract between e_{C1} and e_{U1} is formed, the (personalized) privacy policy $lpp_{ds_{U1}\text{-}dr_{C1}}$ will be saved along with the data of e_{U1}. Therefore, $lpp_{ds_{U1}\text{-}dr_{C1}}$ is not intended for storing the actual data but to reference it.

5. *Transfer:* If e_{C1} transfers the data to e_{C2} the contract formed between those two entities is also converted into a LPL privacy policy $lpp_{ds_{C1}\text{-}dr_{C2}}$. And the existing personalized privacy policy will be added as an underlying privacy policy upp.

$$lpp_{ds_{C1}\text{-}dr_{C2}} = (version, \text{'LPP2'}, lang, ppURI, \{lpp_{ds_{U1}\text{-}dr_{C1}}\}, ds_{C1}, \{p_{U1}\}) \tag{32}$$

This allows a tracking of the data to its origin privacy policy. A (legal) privacy-aware usage is also possible, because each legal usage of data is defined by LPL and can be traced to the first consent between e_{U1} and e_{C1}. This step may be repeated several times if needed.

6. *Usage:* Whichever entity (in this scenario e_{U1}, e_{C1} or e_{C2}) wants to use the data it has to be verified that the entity is authenticated and authorized to query the data. If this is successful the data can be anonymized according to the corresponding purposes.

Summarizing a LPL privacy policy *lpp* should represent and ensure legal privacy policies utilizing the steps *Creation, Negotiation, Pre-Processing, Storage, Transfer* and *Usage*.

5.2 Query-Based Anonymization

In this step, LPL enables a query-based anonymization of the data of the *Data Subject* on purposes which have been consented to and expressed by the LPL privacy policy. Therefore, a query

$$q = \{e_{req}, p_{req}, \widehat{D}_{req}\} \tag{33}$$

is assumed as a request consisting of the following elements:

- e_{req}: The requesting entity. A e_{req} should be first authenticated and authorized to have access to the data.
- p_{req}: The purpose for which the data is requested. Data is only allowed to be used for designated purposes which are either consented to or given by the law.
- D_{req}: The requested data attributes are given to prevent undesirable access.

Hereby, the processes denoted as *Entity-Authentication, Purpose-Authorization, Entity-Authorization, Data-Authorization, Minimum Anonymization* and *Application of Privacy Model* will be conducted in the given order to allow a query-based anonymization (see Fig. 4).

Before the denoted process will be described in detail in the following, supportive data structures will be introduced.

Supportive Data Structures. For processing the LPL privacy policies, additional data structures are assumed available for a *Controller*. These are *Entity-Hierarchy, Entity-Lookup Table* $\widehat{E}_{lookup}$ and *Purpose-Hierarchy* including *Regulated Purposes*, which will be presented in the following. The usage of both *Entity-Hierarchy* and $\widehat{E}_{lookup}$ will be shown in the following sections for *Entity-Authentication* and *Entity-Authorization*. The usage of *Purpose-Hierarchy* and *Regulated Purposes* is shown during *Purpose-Authorization* process.

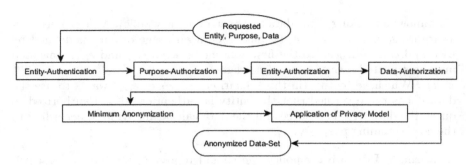

Fig. 4. Processes of the query-based anonymization.

Entity-Hierarchy. The *Entity-Hierarchy* allows to define *Child-Entities* that inherit the rights of the *Parent-Entity* (see Fig. 5). We assume that a privacy policy defines general data recipient roles e.g. a company, but for further control of the processing, which is defined in method descriptions, more fine-grained roles, e.g. a marketing department or even individual employees, have to be defined. This will be enabled by *Entity-Hierarchy*. Only the unique *name*, which specifies roles of e, is needed. This enables the *lpp* creator to define *ds* or *dr*, e.g. 'The user ds_{U1} accepts that the data is used by company dr_{C1} and dr_{C2}.'. If e_{U1} accepts this then dr_{C1} is granted the right to process the data according to the defined purpose. To limit the usage within dr_{C1} to a sub-set of employees, additional entities have to be defined to inherit the rights. It is also possible that a *Child-Entity* inherits from two *Parent-Entities*. This represents the use case when several users allow a company to use their data.

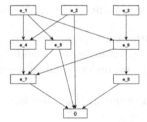

$\widehat{E_{lookup}}$	
name	authorityChain
$name_1$	$authorityChain_1$
$name_2$	$authorityChain_2$
...	...
$name_n$	$authorityChain_n$

Fig. 5. Possible structure of *Entity-Hierarchy*.

Fig. 6. General structure of $\widehat{E}_{lookup}$.

Entity-Lookup Table. The $\widehat{E}_{lookup}$ is the set of all entities (*ds* and *dr*) that exist within all stored *lpp* and additionally all entities that are defined in *Entity-Hierarchy*. Each entry of $\widehat{E}_{lookup}$ has to define the *name* and *authInfo* (see Fig. 6). The *authInfo* resembles the value that is authenticated against, e.g. the

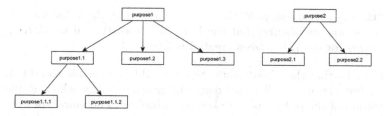

Fig. 7. *Purpose-Hierarchy* showing a possible inheritance hierarchy.

public key if a public/private key authentication is used or the hashed password if a username/password authentication is used. We introduce it to have the possibility to look up entities during the authentication process and potentially other processes without traversing the *Entity-Hierarchy*.

Purpose-Hierarchy and Regulated Purposes. We introduce the *Purpose-Hierarchy* for the *Purpose-Authorization*. The *Purpose-Hierarchy* is a data structure that consists of several trees of purposes. For each purpose it is possible to define *Child-Purposes* that inherit the rights of the *Parent-Purpose* (see Fig. 7). There-fore, only the unique *name* of p is needed which specifies purposes. This enables the *lpp* creator (e.g. dr_{C1}) to define a p in general in the first step, e.g. 'The user accepts that the d_{postal} and d_{salary} is used for '*Marketing*'.', and in the second step the privacy officer can define fine-grained inherent processes matching the requirements of a method description.

We do not assume that it is possible that a *Child-Purpose* inherits from two *Parent-Purposes*.

Additionally, we introduce *Regulated Purposes* that are given by law and regulations and don't have to be described by *lpp*. For example for the basic right for disclosure of confidential information [3, Art. 15] we denote the purpose '[*disclosure*]'. Those purposes allow entities to have access to data based upon the laws and regulations. For example a government agency might have the right to access specific data for a *Regulated Purpose* or a user is allowed to access its own data based upon the introduced *Regulated Purpose* '[*disclosure*]'.

Entity-Authentication. *LPL* will not be restricted to an access control solu-tion by itself but builds upon existing access control methodology that we split into *Entity-Authentication* and *Entity-Authorization*.

Entity-Authentication is necessary to identify an entity e_{req} that requests the *usage* of data. In the following we will focus on the *Authentication* of an entity e against a privacy policy *lpp*. We will show how the previous structures will be used during the LPP life-cycle.

Creation. In the *Creation* step of a privacy policy *lpp* the *dr* entities will be defined. Assuming we have a privacy policy lpp_{raw} from Eq. (22). The purpose p_1 from Eq. (23) allows dr_{C1}, representing the company e_{C1}, to use the data $\widehat{D}_1$. For each dr, the corresponding $publicKey$ will be added as the value for $authInfo$. For the *Creation*, we assume that the key pair, consisting of $publicKey_{C1}$ and $privateKey_{C1}$ for dr_{C1}, has been created beforehand.

Negotiation. During the *Negotiation* step it would be possible to add the ds_{U1} to lpp_{raw}, but the user still could deny the privacy policy which would make the generation of the public/private key pair obsolete. Therefore, the generation of $publicKey_{U1}$ and $privateKey_{U1}$ for ds_{U1} has to be conducted during the *Pre-Processing* step after consent is found.

Pre-Processing. During the *Pre-Processing* step, the set-up for the ds is conducted resulting in $lpp_{ds_{U1}-dr_{C1}}$, whereas the ds_{U1} from Eq. (18) is added to lpp_{raw}. Additionally, for each available e of $lpp_{ds_{U1}-dr_{C1}}$ including the corresponding $publicKey$ will be saved in the $\widehat{E}_{look\text{-}up}$, if not available already. This cannot be done during the *Creation* because the user ds_{U1} has not been specified yet. Therefore, the $\widehat{E}_{look\text{-}up}$ will consist of ds_{U1}, dr_{C1} and dr_{C2}.

$$\widehat{E}_{look-up} = \{ds_{U1}, dr_{C1}, dr_{C2}\} \tag{34}$$

It is important to note that $\widehat{E}_{look\text{-}up}$ may be extended by additional entities, e.g. departments or employees, by the corresponding privacy officer.

Usage. During the *Usage* step an entity e_{req} requests data protected by the privacy policy $lpp_{ds_{U1}-dr_{C1}}$. To ensure that e_{req} is allowed to use the data, it has to be authenticated first. We assume the following scenarii:

Scenario 1: The employee e_{req1} of company C1 requests data concerned by $lpp_{ds_{U1}-dr_{C1}}$.

Scenario 2: The user e_{req2} requests data concerned by $lpp_{ds_{U1}-dr_{C1}}$.

The requesting entities have the following configuration

$$e_{req1} = (\text{`C1'}, classification, privateKey_{C1}, type) \tag{35}$$

$$e_{req2} = (\text{`U1'}, classification, privateKey_{U1}, type) \tag{36}$$

where e_{req1} is an employee using the authentication credentials from company e_{C1} and where e_{req2} represents the user e_{U1} from which the data protected by $lpp_{ds_{U1}-dr_{C1}}$ originates.

In general, we assume the following authentication process for our scenario. The $\widehat{E}_{look\text{-}up}$ is traversed to identify matching entities to the requesting entity e_{req}. We define that two entities match for our scenario if the following is valid.

$$e_{requesting}.name == e.name \tag{37}$$

If a matching entity e_{match} is found then the $publicKey_{match}$ is used to encrypt a *nonce* and sent to e_{req}.

$$encryptedMessage_{match} = encrypt(nonce, publicKey_{match}) \tag{38}$$

To successfully authenticate e_{req} the computed $encryptedMessage_{match}$ has to be decrypted with the $privateKey_{requesting}$ and sent back.

$$decryptedMessage = decrypt(encryptedMessage_{match}, privateKey_{requesting}) \tag{39}$$

The requesting entity will be authenticated if the $decryptedMessage$ equals the $nonce$.

$$message = decrypt(encrypt(message, publicKey), privateKey) \tag{40}$$

With this authentication mechanism we will describe the given scenarii. For scenario 1, the requesting entity is e_{req1}. In $\widehat{E}_{look\text{-}up}$ the matching entity is dr_{C1} with both $name$ values are 'C1'. In the next step it will be evaluated if e_{req1} can authenticate as the dr_{C1} utilizing the public/private key authentication shown before

$$nonce = decrypt(encrypt(nonce, publicKey_{C1}), privateKey_{C1}) \tag{41}$$

which results in the authentication success for scenario 1.

In scenario 2, the requesting entity is e_{req2}. In $\widehat{E}_{look\text{-}up}$ the matching entity is ds_{U1}. Therefore, the authentication will be executed with $publicKey_{U1}$ from ds_{U1} and $privateKey_{U1}$ from e_{req2}.

$$nonce = decrypt(encrypt(nonce, publicKey_{U1}), privateKey_{U1}) \tag{42}$$

This successful authentication will result in the authorization of e_{U1} as ds of $lpp_{ds_{U1}\text{-}dr_{C1}}$, which allows the requesting entity to have all processing-rights authorized for the described data in $lpp_{ds_{U1}\text{-}dr_{C1}}$ as shown in the following.

In general, the authentication process for an entity requires the name of the requesting entity ($entityName$) as well as a $secret$ to verify the identity. The calculation of the $secret$ depends on the individual implementation of the authentication process which is based upon the $authInfo$. If a private/public key pair is used like in the aforementioned scenarii then the secret will be the $decryptedMessage$ that has to be verified against the $nonce$. Additional steps for setting up the $encryptedMessage$ and transferring it to the e_{req} would be necessary.

Purpose-Authorization. In the scenarii for *Purpose-Authorization*, we assume that the requesting entity e_{req} is authenticated and therefore only focus on the verification of the purpose p_{req} given in the query. The query is rejected if the purpose is invalid.

We assume that a purpose can have *Child-Purposes*, which are stored in the *Purpose-Hierarchy*. We assume the set-up of $lpp_{ds_{U1}\text{-}dr_{C1}}$ from Eq. (31). Moreover, we assume that the purpose '*Marketing*', as well as its child-purpose '*Newsletter*' and the purpose '*Development*' are available in *Purpose-Hierarchy*, next to the *Regulated Purposes* given by law and regulations, which we restrict to *[disclosure]* in the *Purpose-Hierarchy* (see Fig. 8). Therefore, we will describe the following scenarii for our set-up:

Scenario 1: Entity e_{req1} requesting the $\widehat{D}_1$ protected by $lpp_{ds_{U1}\text{-}dr_{C1}}$ for the purpose '*Newsletter*'.

Scenario 2: Entity e_{req2} requesting the $\widehat{D}_1$ protected by $lpp_{ds_{U1}\text{-}dr_{C1}}$ for the purpose '*[disclosure]*'.

In scenario 1, the entity e_{req1} requests data for the purpose of '*Newsletter*'. The data is protected by $lpp_{ds_{U1}\text{-}dr_{C1}}$ containing only purpose p_{U1} which *name* is '*Marketing*' defining the authorized purpose. Because we consider a *Purpose-Hierarchy*, where every *Child-Purpose* of the authorized purpose is also authorized, a set of *Authorized Purposes* has to be generated. Therefore, the requested purpose '*Newsletter*' will be identified in the *Purpose-Hierarchy* (see Fig. 8) and all its *Parent-Purposes* as well as the purpose itself will be returned {'Marketing', 'Newsletter'}. Then, the purpose is considered as an authorized purpose as its name behaves to the calculated purpose set.

In scenario 2, the entity e_{req2} requests data for the '*[disclosure]*' purpose. The corresponding set of *Authorized Purposes* consists therefore of {[disclosure]}. The purpose cannot be found in $lpp_{ds_{U1}\text{-}dr_{C1}}$, because this is a *Regulated Purpose* defining a special case and a new purpose has to be crafted for it during runtime automatically. For '*[disclosure]*' a new purpose $p_{disclosure}$ will be computed

$$p_{disclosure} = (\text{'[disclosure]'}, optOut, required, descr, \{dr_{U1}\}, \tag{43}$$
$$r, pm, \{d'_{postal}, d'_{salary}\})$$

$$dr_{U1} = (\text{'U1'}, \text{'Person'}, publicKey_{U1}, \text{'DataRecipient'}) \tag{44}$$

containing all d of the corresponding $lpp_{ds_{U1}\text{-}dr_{C1}}$ and the ds_{U1} is transformed to the dr of $p_{disclosure}$. Hereby d'_{postal} and d'_{salary} will be created by removing the respective am if existent. The authorization of e_{req2} will be executed according to the *Entity-Authorization*. Therefore, e_{req2} is requesting an authorized purpose.

In general, the authorization process for a purpose requires the *name* of the purpose and the corresponding lpp (see Listing 1). The authorization is successful if the *name* of the requested purpose or any of the corresponding *Authorized Entities* matches any of the p of the lpp or if a *Regulated Purpose* is requested.

Entity-Authorization. In the following, we will focus on the *Entity-Authorization*. An entity e_{req} is authorized to use the data if the dr or ds *name* of the lpp matches the *name* of e_{req}, whereas *name* can either specify a role or a specific user. We assume the employees e_{req3} and e_{req4} respectively from the marketing departments M1 and M2 of company C1 and C2.

$$e_{req3} = (\text{'M1'}, \text{'Person'}, authInfo, \text{'DataRecipient'}) \tag{45}$$
$$e_{req4} = (\text{'M2'}, \text{'Person'}, authInfo, \text{'DataRecipient'}) \tag{46}$$

We assume the set-up of $lpp_{ds_{U1}\text{-}dr_{C1}}$ from Eq. (31). User e_{U1} gave its consent on the usage of its personal information for the purpose of '*Marketing*' for

```
1 P authorizePurpose(purposeName, lpp):
2
3      //initialize authorizedPurposes
4      authorizedPurposes = {};
5
6      //receive set of possible purposes
7      possiblePurposes = purposeHierarchy.getParentPurposes(purposeName);
8
9      if possiblePurposes != null
10         //verify if purpose matches at least one p of lpp
11         for possibleP : possiblePurposes
12
13             switch (possibleP)
14                 //for each regulated purpose a individual case
15                 case '[disclosure]'
16                     authorizedPurposes.add(createDisclosureP(lpp));
17
18                 //add purpose if name match
19                 default
20                     for p : lpp.P
21                         if match(possibleP, p.name)
22                             authorizedPurposes.add(p);
23
24     return authorizedPurposes;
```

Listing 1. Pseudocode describing the authorization of purposes of an *lpp* utilizing *Purpose-Hierarchy*. The special cases of *Regulated Purposes* are shown exemplary for *'disclosure'*. The *Entity-Hierarchy* is assumed to be accessible within the method.

company e_{C1}, which is encoded in $lpp_{ds_{U1}\text{-}dr_{C1}}$. No further agreements exist between user e_{U1} and company e_{C1}. Company e_{C1} has a marketing department '*M1*'. Company e_{C2} has a marketing department '*M2*'. Company e_{C1} grants e_{M2} from company e_{C2} access to the e_{U1} data for the purpose of '*Marketing*' encoded in $lpp_{ds_{C1}\text{-}dr_{M2}}$.

$$e_{M2} = (\text{'M2'}, \text{'Legal Entity'}, authInfo, \text{'DataRecipient'}) \tag{47}$$

$$p_{C1} = (\text{'Marketing'}, optOut, required, descr, \{e_{M2}\}, r, pm, \widehat{D}) \tag{48}$$

$$lpp'_{ds_{C1}\text{-}dr_{M2}} = (version, name, lang, ppURI, lpp_{ds_{U1}\text{-}dr_{C1}}, ds_{C1}, \{p_{C1}\}) \tag{49}$$

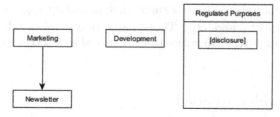

Fig. 8. *Purpose-Hierarchy* for the *Purpose-Authorization* scenarii. The *Regulated Purposes* are separated for better understanding.

Company e_{C1} and company e_{C2} exchange their data in a privacy conform way. We will present the following scenarii:

Scenario 1: The employee e_{req3} of the marketing department '$M1$' requests personal information of user e_{U1} for the purpose of '$Marketing$'.

Scenario 2: The employee e_{req4} of the marketing department '$M2$' requests personal information of user e_{U1} for the purpose of '$Marketing$'.

Scenario 3: The user e_{U1} makes use of him being entitled to the disclosure of confidential information [3, Art. 15], which is the basis for several additional interests, towards company e_{C1}. Therefore, e_{req2} requests its personal information for the purpose of '[$disclosure$]'.

For authorization we assume a role-based access control (RBAC) system [43] with the *roles* in Fig. 9. The *permission* is provided by the LPL privacy policy.

In scenario 1, the employee e_{req3} has the role '$M1$'. Based upon the (underlying) privacy policy $lpp_{ds_{U1}\text{-}dr_{C1}}$ the role '$C1$' is granted to use the data for '$Marketing$'. '$M1$' is a child-role of '$C1$' inheriting the permission to use the personal data of e_{U1}. Furthermore, the stated purpose in the LPL privacy policy '$Marketing$' matches the purpose of the requester. This concludes that e_{req3} is authorized to access the personal information of e_{U1} in scenario 1.

In scenario 2, an employee e_{req4} of company e_{C2} requests to access the personal data of e_{U1}. The value '$M2$' e_{req4} matches the data recipient dr of p_{C1}. The purpose '$Marketing$' matches the purpose defined in p_{C1} and therefore e_{req4} is authorized to access the data.

In scenario 3, the requesting entity e_{req2} matches the *DataSource*-element and therefore e_{U1} is authorized to request the personal data of itself.

In general, the authorization process for an entity requires the *name* of the e_{req} as well as the purpose for which it should be verified against (see Listing 2). The authorization is successful if the *name* of e_{req} or any of the corresponding *Parent-Entities* matches any of the dr of the purpose. This process has to be conducted after the *Purpose-Authorization* has been executed to consider special cases that are defined by the law and regulations as shown in scenario 3.

Data-Authorization. We assume that e_{req} is authenticated, authorized and uses an authorized purpose for the following scenarii. Hereby, the requested data $\widehat{D}_{req}$ has to be verified against the described data within the authorized purposes. If the verification is not successful, then the query will be rejected.

For the following scenarii we assume that an entity e_{req1} is querying different sets of data $\widehat{D}_{q1}$ and $\widehat{D}_{q2}$. Those requests are validated against p_{U1} of $lpp_{ds_{U1}\text{-}dr_{C1}}$. It is important to notice that p_1 only allows access to d_{postal}, d_{salary} and d_{age}.

$$\widehat{D}_{q1} = \{d_{postal}, d_{salary}\} \tag{50}$$

$$\widehat{D}_{q2} = \{d_{postal}, d_{salary}, d_{age}\} \tag{51}$$

$$d_{age} = (\text{'age'}, dGroup, dType, required, descr, pGroup, \emptyset) \tag{52}$$

```
 1 boolean authorizeEntity(entityName, p):
 2
 3     //receive set of authorized entities
 4     authorizedEntities = entityHierarchy.getParentEntities(entityName);
 5
 6     if authorizedEntitites != null
 7         //verify if entity matches at least one dr of p
 8         for dr : p.DR
 9             if match(entityName, dr.name)
10                 return true;
11
12     return false;
```

Listing 2. Pseudocode describing the authorization of an requesting entity e against a single *lpp* utilizing *Entity-Hierarchy*. The *Entity-Hierarchy* is assumed to be accessible within the method.

Additionally, we assume that the value of the *name* identifies a data-field and no additional matching between the *name* and the stored data-fields is necessary. This may change in a real world scenario but will not change basic behaviour that will be described in the following scenarii for this set-up:

Scenario 1: Entity e_{req1} requesting the $\widehat{D}_{q1}$ from p_{U1} of $lpp_{ds_{U1}-dr_{C1}}$.

Scenario 2: Entity e_{req1} requesting the $\widehat{D}_{q2}$ from p_{U1} of $lpp_{ds_{U1}-dr_{C1}}$.

In scenario 1, the entity e_{req1} requests $\widehat{D}_{q1}$ for the purpose of '*Marketing*'. In this scenario the requested set $\widehat{D}_{q1}$ is a sub-set of the data-set $\widehat{D}_1$ defined in p_{U1}. This means that all requested data-fields are defined in the authorized purpose and therefore the usage of the data is authorized. In scenario 2, the entity e_{req1} requests $\widehat{D}_{q2}$ for the purpose of '*Marketing*'. In this scenario, the requested set $\widehat{D}_{q2}$ is evaluated against the data-set $\widehat{D}_1$ defined in p_1. Each in $\widehat{D}_{q2}$ defined entry has to be also defined in $\widehat{D}_1$ and return an invalid result as the requested data '*age*' is not a member of $\widehat{D}_1$. Therefore, the usage of the data '*age*' is not authorized and the whole query will be rejected.

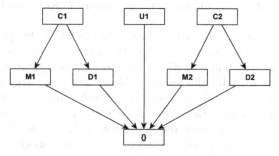

Fig. 9. Example roles for the *Entity-Authorization* scenarii. The {} denotes a default role without any processing-rights.

```
1 D authorizeData(requestedD, P):
2
3     //initialize authorizedData
4     authorizedData = {};
5
6     //check for each requested data if it is authorized
7     for d : requestedD
8         //several authorized purposes are possible
9         for p : P
10            for dAuthorized : p.D
11                if match(d.name, dAuthorized.name)
12                    authorizedData.add(d);
13
14    return authorizedData;
```

Listing 3. Pseudocode describing the authorization of data from *Authorized Purposes*. The set of requested is assumed to be computed from a parsed query.

In general, the authorization process for data requires the *name* of the data of the corresponding purpose p (see Listing 3). The authorization is successful if the *name* of the requested data matches any of d from any *AuthorizedPurpose*. This process has to be conducted after the *Purpose-Authorization* has been executed.

Minimum Anonymization. The *AnonymizationMethod*-element is introduced to specify (personalized) privacy settings for each *Data*-element. Only the relevant elements and attributes will be described for better understanding in the following three steps - *Negotiation*, *Pre-Processing* and *Usage*. We denote the anonymization of the data during the *Usage* step as *Minimum Anonymization* which is conducted after the *Data-Authorization*. We assume the following scenario (see Fig. 10).

The personal data $\widehat{D}$ of an user e_{U1} is requested by a company e_{C1}. The data is requested for the purpose '*Marketing*'. The corresponding purpose protecting the data is p_{U1} which is described by $lpp_{ds_{U1}-dr_{C1}}$ from Eq. (31). We focus only on the personal data for *postal-code* d_{postal} with the value of '94032' (for Passau in Germany). The configuration of am_1 from Eq. (26) describes the anonymization method '*Suppression*' with the replacement character '*' starting from the last character '*backwards*'. The minimum and maximum suppression levels are given.

Negotiation. In the *Negotiation* step it has to be verified if the, possibly from the user personalized, privacy policy is valid. The privacy policy is valid if, among to other conditions, the value of *Minimum Level* is not greater than the value of *Maximum Level*. Initially the value of *Maximum Level* and *Minimum Level* will be defined by e_{U1}, whereas *Maximum Level* defines the maximal usable anonymization for e_{U1} and the value of *Minimum Level* is an initial recommended proposal for the privacy requirements. This asserts the validity of the privacy policy before it is stored. We assume that the integrity of the value of *Maximum Level* is preserved.

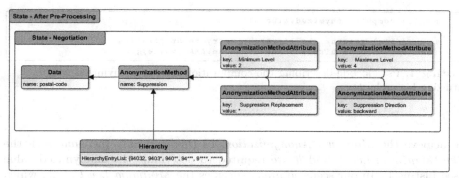

Fig. 10. Relevant elements and attributes for the anonymization scenario in the state during the *Negotiation* and after the *Pre-Processing* step.

Pre-Processing. In the *Pre-Processing* step a *Hierarchy*-element h will be created for each anonymization method am. This step is conducted after the *consent* of the user for the privacy policy is given and before the data and the privacy policy are stored. In the *AnonymizationMethod*-element the *name* specifies the method applied on the data value. In our scenario, we assume that the anonymization method chosen by the user is *Suppression*. The set of *AnonymizationMethodAttributes* ama is utilized for this process. The ama_1 describes the character that is used for the replacement during the suppression, which is '$*$' in this scenario. The ama_2 describes the variation of the anonymization method. Thus, current scenario suppresses the value with '$*$' starting with the end of the *postal-code*. According to this configuration the *Hierarchy*-element will be created, which contains an ordered list of values, and added to the am. In our scenario the hierarchy h_1 for the postal-code '94032' will contain $\widehat{\mathbf{HE_1}}$.

$$\widehat{\mathbf{HE_1}} = \{\text{'94032', '9403*', '940**', '94***', '9****', '*****'}\} \tag{53}$$

We denote that the first value is at *Level* '0' and the last element, in this case, is at *Level* '5'. This *Level* will be referred to by the ama with *Minimum Level* and *Maximum Level*. The hierarchy h_1 will be added to am_1' replacing am_1.

$$am_1' = (\text{'Suppression'}, \{ama_1, ama_2, ama_3, ama_4\}, h_1) \tag{54}$$

$$h_1 = (\widehat{\mathbf{HE_1}}) \tag{55}$$

Therefore, the h_1 holds all possible anonymized values for the data value. The values are ordered in an hierarchical way from the least to most anonymized value.

Usage. In the *Usage* step we assume e_{req1} is requesting the data of e_{U1} for the purpose '*Marketing*'. The corresponding p_1 for the request will be processed and therefore the defined anonymization method am_1' has to be applied

```
1 value computeAnonymizedValue(d):
2
3     //h(n) returns the hierarchy entry he at position 'n'
4     return d.am.h(selectMinLevel(d.am.AMA).value);
```

Listing 4. Pseudocode describing the computation for the anonymized value utilizing hierarchy h.

to achieve the *Minimum Anonymization*. In this step only the *ama* with the key '*MinimumLevel*' and h are required to determine the anonymized value (see Listing 4). In our scenario ama_3 specifies the *Minimum Level* of '2', which means that the value on *Level* '2' of h_1 has to be selected. This results in the anonymized value of '940 ∗ ∗' for the *postal-code*.

Application of Privacy Model. The functionality of the *PrivacyModel*-element and its corresponding *PrivacyModelAttribute*-element will be explained in the following.

In general, a privacy model describes the probability of a record in a data-set to be identified or de-anonymized [41]. For each record in a data-set, a privacy model can be defined utilizing *pm* and *pma* of LPL. The *pm* will be processed after the *Minimal Anonymization* has been conducted during the *Usage* step. To avoid computational overhead, we decided to compute the minimum required privacy model pm_{min} and apply it on the data-set. We specify the minimum required privacy model pm_{min} as the privacy model with the highest privacy requirements to guarantee that no initially given privacy constraints are violated. Therefore, the set of all defined privacy models has to be substituted. We decided to consider the attacks which are mitigated by the privacy models for a classification. Table 3 represents such a classification. Based upon this classification, a rule-set for minimizing the used privacy models can be created beforehand which we denote as *Privacy Model Substitution Table*. According to the classification, l-Diversity, which is the direct successor to k-Anonymity [9], covers all attacks of k-Anonymity, namely *Record Linkage* and *Attribute Linkage* [41]. Therefore, the set of privacy models 'k_{i1}-*Anonymity*' and 'l_{i2}-*Diversity*' will be substituted to '$l_{r1} - Diversity$'. The value for the parameter 'l_{r1}' of the resulting 'l_{r1}-*Diversity*' has to be calculated that it fulfills the privacy requirements of all prior privacy models. Both values can be treated equivalent because l-Diversity and k-Anonymity use similar definitions for the privacy. Therefore, 'l_{r1}' will have the maximum value of both 'k_{i1}' and 'l_{i2}' resulting in '($l_{r1}, max(k_{i1}, l_{i2})$)'. According to the properties of t-Closeness, the minimum of the parameters will be used for the substitution of two t-Closeness privacy models [10]. The substitution with k-Anonymity or l-Diversity results in no reduction, because t-Closeness does not cover all the attack models of the other privacy models. Additional rules will be created by this scheme resulting in the *Privacy Model Substitution Table*.

Table 3. Excerpt of mitigated attack models by privacy models [41].

Privacy model	Attack model			
	Record linkage	Attribute linkage	Table linkage	Probabilistic attack
k-Anonymity	x			
l-Diversity	x	x		
t-Closeness		x		x

Table 4. *Privacy Model Substitution Table* for the scenarii used in Sect. 5.2.

Privacy model set	Substitution privacy model	Substitution privacy model attribute
$\{k_{i1}\text{-}Anonymity, k_{i2}\text{-}Anonymity\}$	$\{k_{r1}\text{-}Anonymity\}$	$\{(k_{r1}, max(k_{i1}, k_{i2}))\}$
$\{l_{i1}\text{-}Diversity, l_2\text{-}Diversity\}$	$\{l_{r1}\text{-}Diversity\}$	$\{(l_{r1}, max(l_{i1}, l_{i2}))\}$
$\{t_{i1}\text{-}Closeness, t_{i2}\text{-}Closeness\}$	$\{t_{r1}\text{-}Closeness\}$	$\{(t_{r1}, min(t_{i1}, t_{i2}))\}$
$\{k_{i1}\text{-}Anonymity, l_{i2}\text{-}Diversity\}$	$\{l_{r1}\text{-}Diversity\}$	$\{(l_{r1}, max(k_{i1}, l_{i2}))\}$
$\{k_{i1}\text{-}Anonymity, t_{i2}\text{-}Closeness\}$	$\{k_{r1}\text{-}Anonymity\}, \{t_{r2}\text{-}Closeness\}$	$\{(k_{r1}, k_{i1}), (t_{r2}, t_{i2})\}$
$\{l_{i1}\text{-}Diversity, t_{i2}\text{-}Closeness\}$	$\{l_{r1}\text{-}Diversity\} \{t_{r2}\text{-}Closeness\}$	$\{(l_{r1}, l_{i1}), (t_{r2}, t_{i2})\}$

Note that it is possible that another privacy model exists which covers all attack models which would be more suitable for a substitution, but we limit our example only on k-Anonymity, l-Diversity and t-Closeness.

The computation of pm_{min} and the application on the data-set will be described in the following. For the scenarii we use the *Privacy Model Substitution Table* shown in Table 4. We assume records for e_{E1}, e_{E2}, e_{E3} and e_{E4} representing entries in a database table. For each record we assume that the record contains the postal-code d_{postal} and the salary per year d_{salary}. The values for each record are summarized in Table 5 for each e.

Table 5. Values for the d_{postal} and d_{salary} for each record.

Entity	Postal-code	Salary
e_{E1}	94032	36.000
e_{E2}	94032	45.000
e_{E3}	94034	38.000
e_{E4}	94032	45.000

We assume the privacy models 2-Anonymity pm_1, 3-Anonymity pm_2 and 2-Diversity pm_3.

$$pm_1 = (\text{'k-Anonymity'}, \{pma_1\}) \tag{56}$$

$$pma_1 = (\text{'k'}, \text{'2'}) \tag{57}$$

$$pm_2 = (\text{'k-Anonymity'}, \{pma_2\}) \tag{58}$$

$$pma_2 = (\text{'k'}, \text{'3'}) \tag{59}$$

$$pm_3 = (\text{'l-Diversity'}, \{pma_3\}) \tag{60}$$

$$pma_3 = (\text{'l'}, \text{'2'}) \tag{61}$$

Only d_{postal}, which is classified as 'QID', will be considered in the anonymization process of k-Anonymity [8]. The specified d and pm are combined in the corresponding p individually for each record.

$$p_{E1} = (\text{'p'}, optOut, required, descr, \widehat{DR}, r, pm_1, \{d_{postal}, d_{salary}\}) \tag{62}$$

$$p_{E2} = (\text{'p'}, optOut, required, descr, \widehat{DR}, r, pm_1, \{d_{postal}, d_{salary}\}) \tag{63}$$

$$p_{E3} = (\text{'p'}, optOut, required, descr, \widehat{DR}, r, pm_2, \{d_{postal}, d_{salary}\}) \tag{64}$$

$$p_{E4} = (\text{'p'}, optOut, required, descr, \widehat{DR}, r, pm_3, \{d_{postal}, d_{salary}\}) \tag{65}$$

Explicit and non-sensitive attributes have been omitted for the following scenarii. For e_{E1} and e_{E2} the privacy model 2-Anonymity pm_1 is defined, for e_{E3} 3-Anonymity pm_2 is defined and for e_{E4} 2-Diversity pm_3 is defined.

Scenario 1: Data of e_{E1}, e_{E2} and e_{E3} are queried, the corresponding purposes are p_{E1}, p_{E2} and p_{E3}.

Scenario 2: Data of e_{E1}, e_{E3} and e_{E4} are queried, the corresponding purposes are p_{E1}, p_{E3} and p_{E4}.

We assume the same purpose 'p' for each query and will describe the computation of pm_{min} and the outcome for each scenario in the following.

In scenario 1, the data of the entities e_{E1}, e_{E2} and e_{E3} is queried. The *name* of the corresponding privacy models in pm_1, pm_1 and pm_2 are all the same. Therefore, the *value* of the corresponding attributes pma_1, pma_1 and pma_2 have to be compared, which show different configurations. The computation of the pm_{min1} results in the value '3' for k, because no additional conflicts occur. The process will be executed in two steps. First pm_1 and pm_1 will be substituted the resulting pm will then be substituted with pm_2 to compute pm_{min1} finally. Therefore, for scenario 1 the valid privacy model for the data-set is 3-Anonymity.

$$pm_{min1} = pm_2 = (\text{'k-Anonymity'}, \{pma_2\}) \tag{66}$$

Considering the corresponding values of Table 5 the data-set will be anonymized. The initial table T will be anonymized to table T', whereas the postal-code will by suppressed to '9403*' for all records to achieve 3-Anonymity (see Table 6).

```
 1 PM calculateMinimumPM(PM):
 2
 3     resultPM = {};
 4     iteratorPM = PM.iterator();
 5
 6     //initialize
 7     if resultPM.isEmpty()
 8         resultPM.add(iteratorPM.next());
 9
10     while iteratorPM.hasNext()
11         tempResultPM = resultPM;
12         pm = iterator.next();
13
14         for pm' : resultPM
15             //subsitutePM utilizes PrivacyModelSubstitutionTable
16             minPM = substitutePM(pm',pm);
17
18             //replace pm' in resultPM if necessary
19             if not match(pm', minPM)
20                 tempResultPM.remove(pm');
21                 tempResultPM.add(minPM);
22
23         resultPM = tempResultPM;
24
25     return resultPM;
```

Listing 5. Pseudocode describing possible algorithm to select privacy models.

In scenario 2, the entities e_{E1}, e_{E3} and e_{E4} are queried. The *name* of the corresponding privacy models pm_1, pm_2 and pm_3 differ in this scenario. There exists a conflict between the privacy models '*k-Anonymity*' and '*l-Diversity*'. For the computation of the pm_{min2} both the correct privacy model and the corresponding value has to be determined by substituting first pm_1 and pm_2 and then substitute the result with pm_3. According to the *Privacy Model Substitution Table* the *name* for pm_{min2} will be '*l-Diversity*' with the *value* '3' for '*l*'.

$$pm_{min2} = (\text{'l-Diversity'}, \{pma_{min2}\}) \qquad (67)$$
$$pma_{min2} = (\text{'l'}, \text{'3'}) \qquad (68)$$

Considering the corresponding values of Table 5 the data-set will be anonymized. The initial table T will be anonymized to table T' (see Table 7). We assume for this scenario that the salary per year will be generalized to '50.000' and the postal-code will be suppressed to '9403*' to match the conditions of 3-Diversity. In the scenarii we only showed the examples resulting in one privacy model. But it is also possible that the result contains several privacy models after the substitution has been conducted (see Listing 5).

5.3   Provenance

The *UnderlyingPrivacyPolicy* is introduced in LPL for *Privacy Policy Provenance*. This means that LPL enables to distinguish between different privacy policies and their origin. In this example we will show how the provenance is

Table 6. Transformation of table T to table T' for the given k-Anonymity privacy model. Column *Postal-code* is a *QID* and *Salary* is a *Sensitive Attribute*.

User	Table T		Privacy model	3-anonymous table T'	
	Postal-code	Salary		Postal-code	Salary
U1	94032	36.000	2-Anonymity	9403*	36.000
U2	94032	45.000	2-Anonymity	9403*	45.000
U3	94034	38.000	3-Anonymity	9403*	38.000

Table 7. Transformation of table T to table T' for the given k-Anonymity and l-Diversity privacy model. Column *Postal-code* is a *QID* and *Salary* is a *Sensitive Attribute*.

User	Table T		Privacy model	3-diverse table T'	
	Postal-code	Salary		Postal-code	Salary
U1	94032	36.000	2-Anonymity	9403*	<50.000
U3	94034	38.000	3-Anonymity	9403*	<50.000
U4	94032	45.000	2-Diversity	9403*	<50.000

preserved during the *Transfer* and *Usage* step (see Fig. 11). Assuming we have the scenario based on $lpp_{ds_{U1}-dr_{C1}}$ and $lpp_{ds_{C1}-dr_{C2}}$. Hereby, ds_{U1} agrees that the d_{postal} will be used by dr_{C1} and dr_{C2} for the purpose of '*Marketing*'.

Transfer. After the user e_{U1} has agreed on providing the *postal-code* to e_{C1} under $lpp_{ds_{U1}-dr_{C1}}$, e_{C1} can form a contract with e_{C2} for outsourcing the '*Marketing*' task creating $lpp_{ds_{C1}-dr_{C2}}$. Therefore, e_{C1} ensures the correct usage of the personal data by transferring the corresponding privacy policy with the personal data to e_{C2}. To ensure the provenance of the personal data, the privacy policy of the data-source will be added to the privacy policy of the dr. This represents the *Transfer* step, which can be repeated several times.

Usage. We will demonstrate following scenarii after the transfer of the data from e_{U1} to e_{C1} and from e_{C1} to e_{C2} has been executed. Hereby e_{req5} representing e_{C2} and e_{req2}, the original data source, will be compared as requesting entities as follows, demonstrating *Provenance*.

$$e_{req5} = (\text{'C2'}, classification, privateKey_{C2}, type) \tag{69}$$

Scenario 1: Request of d_{postal} from data-warehouse e_{E3} by e_{req5} for '*Marketing*'.

Scenario 2: Request of d_{postal} from data-warehouse e_{E3} by e_{req2} for '[*disclosure*]'.

We show for each scenario if the request is successful and which source for the data can be identified. Therefore, we assume the *Entity-Authentication*,

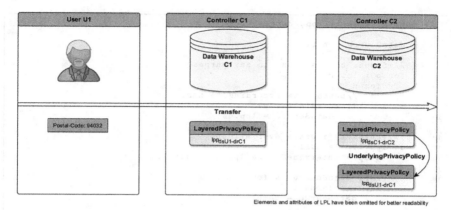

Fig. 11. Relevant elements and attributes for the layered privacy policy scenario visualizing the different privacy policy layers when the data is transferred.

Purpose-Authorization, *Entity-Authorization* and *Data-Authorization* have been conducted successfully.

In scenario 1, the $lpp_{ds_{C1}-dr_{C2}}$ has to be considered. *UnderlyingPrivacyPolicies* are considered successively. For *lpp* the purpose '*Marketing*' will be authorized successfully according to p_{U1}. The request will be successful and the data will be anonymized to the value '940 ∗ ∗' according to am_1. For $lpp_{ds_{C1}-dr_{C2}}$, it is possible to identify e_{U1} as source when the *UnderlyingPrivacyPolicies* are traversed utilizing the algorithm from Listing 6.

For scenario 2, the purpose '[*disclosure*]' will be authorized successfully as a *Regulated Purpose* and $p_{disclosure}$ will be created. Due to the missing *am* of d_{postal} the value '94032' will be returned. The source for the data is identified as the same like in scenario 1. For scenario 2, the *ds* could be identified successfully despite the original data has been transferred several times already and the original value '94032' was returned for e_{req2}. The *UnderlyingPrivacyPolicies* of LPL therefore enable the *Provenance* for the data source.

In general, to determine the source of a data in LPP it is necessary to firstly identify the data by the *name* within a purpose (see Listing 6). If the *lpp* has no *upp* then the *ds* of *lpp* is the source. If *upp* is available it has to be checked for all *p* of it, if the *name* is contained. If so the process is repeated till no *p* with *name* is found or no *upp* is available.

5.4 Retention

The *Retention*-element r provides the information when the data for a specific purpose p has to be deleted. With *Retention* a planned deletion of data is denoted, which has to be differentiated from an action-based deletion, e.g. like it is denoted by the right to erasure [3, Art. 17 No. 1 (b)] [3, Art. 17 No. 1 (c)] in which the user actively withdraws or objects. The retention process is executed during the *Usage* step.

```
1 e determineSource(lpp, data):
2
3     dataSource = null;
4
5     //if data can be found in any purpose
6     for p : lpp.P
7         for d : p.D
8             //match data according to name
9             if match(data, d)
10                dataSource = lpp.ds;
11
12     //recursivly iterate over all upp
13     if lpp.upp != null
14         temp = determineSource(lpp.upp, data);
15
16         //if dataSource is found
17         if temp != null
18             dataSource = temp;
19
20
21     return dataSource;
```

Listing 6. Pseudocode describing possible algorithm to determine the source of data.

There are three basic options that are used to define when the data has to be deleted for a specific purpose.

For the *type Indefinite* there is no designation point in time for the deletion of the data, so the data will not be deleted after a specific time.

$$r_1 = (\text{'Indefinite'}, \emptyset) \tag{70}$$

For the *type AfterPurpose* the deletion of the data depends on the completion of the purpose p itself, which will have to be managed separately within an encapsulating framework.

$$r_2 = (\text{'AfterPurpose'}, \text{'3 months'}) \tag{71}$$

After the purpose the data for this purpose has to be deleted within "3 *months*".

The last *type* is *FixedDate* which defines exactly the deletion.

$$r_3 = (\text{'FixedDate'}, \text{'01.01.2018'}) \tag{72}$$

Defining that on the *01.01.2018* the data of the corresponding purpose has to be deleted. The way the processing of data deletion based upon a rule r differs. One possibility is an automatic data deletion system. Within the system the data will be deleted exactly at the point in time when the r of the privacy policy defines it. The deletion checking has hereby to be done regularly. Further research on the requirements of an automatic data deletion system based on LPL is part of future work.

6   Conclusion and Future Works

This paper presents LPL, a privacy language that takes into account both legal and privacy-preserving requirements. After deriving the main objectives, we have given a formal definition of our privacy language. Later on, we describe the life-cycle of LPL as well as a usage pattern for *Query-based Anonymization* utilizing *Entity-Authentication, Purpose-Authorization, Entity-Authorization, Data-Authorization, Minimum Anonymization* and *Application of Privacy Model*. Additionally, we outline how *Provenance, Retention* is enabled. LPL does not cover all privacy aspects, which are partially already discussed by other works. Furthermore, LPL is an extensible language and can easily let new security and privacy concepts be integrated.

We consider LPL as a work in progress which we will extend in future works, hereby we will cover additional aspects of the GDPR by LPL.

First of all, there is an ongoing implementation work of LPL behind which we aim to validate experimentally a privacy-preserving framework based on LPL to allow an automatic query-based anonymization for data-storages, like data-warehouses. Not only anonymization, but also pseudonymization will be covered in future work and the impact of personal privacy and privacy models on privacy, utility and the performance will be evaluated.

To support privacy-aware applications (e.g. web-applications), which are based on such data-storages, it is necessary to optimize the response time for a query, by minimizing the computational overhead of LPL, so that common usage of applications is not hindered by privacy. We project a set of optimizations along the process steps. As examples we could cite the calculation of the minimum required privacy model or the authorization against every purpose for a requesting entity, because in both cases the execution time depends on the amount of processed privacy policies. Additionally we will extend our *Query-based Anonymization* to consider sequential queries and releases to avoid a privacy breach, were we will also consider logical interference.

Furthermore, we project to implement an user-friendly interface that enables the *Data Subject* to express simply its consent and have a fast overview over the privacy policy. Also personal privacy should be facilitated, allowing the user to refuse predefined purposes or adjust the privacy settings.

We assume that especially the combination of personal privacy and privacy models will influence the utility of the resulting data-set. Hereby, different factors, like the amount of personalized privacy policies or the properties of the data-set, will be investigated with the aim to identify all factors influencing the trade-off between privacy and utility.

Additionally, diverse *Regulated Purposes* have to be identified and implemented by our privacy-preserving framework to support the *Data Subject Rights* of the *GDPR*. On one hand this may relieve the *Controller* from burdensome manual responses and on the other hand this allows the *Data Subjects* to rely on a lawful execution of their requests.

Another future work concerns conflict detection between different privacy policies expressed in different LPL files. Those might occur during the transfer and aggregation of distinct data sources. Assuming the data sources origin from conflicting legal spaces, further investigations for the resolution of conflicts in terms of both legal and technical requirements and capabilities have to be executed. It is imaginable that *Provenance* has to be omitted in such a scenario for the sake of privacy. Therefore, scenarii with trusted and untrusted *Controllers* have to be considered in future works to assess under which circumstances *Provenance* can be provided.

Concluding we want to state that there is a various amount of open challenges in the field of privacy that arise from the GDPR from which LPL focuses on enforceable privacy policies.

References

1. Cranor, L.F., Arjula, M., Guduru, P.: Use of a P3P user agent by early adopters. In: Proceedings of the 2002 ACM Workshop on Privacy in the Electronic Society, WPES 2002, pp. 1–10. ACM, New York (2002)
2. Iyilade, J., Vassileva, J.: P2U: a privacy policy specification language for secondary data sharing and usage. In: Proceedings of IEEE Security and Privacy Workshops, pp. 18–22, May 2014
3. Council of the European Union: General data protection regulation, April 2016. Regulation (EU) 2016 of the European Parliament and of the Council of on the protection of natural persons with regard to the processing of personal data and on the free movement of such data, and repealing Directive 95/46/EC
4. Tsormpatzoudi, P., Berendt, B., Coudert, F.: Privacy by design: from research and policy to practice – the challenge of multi-disciplinarity. In: Berendt, B., Engel, T., Ikonomou, D., Le Métayer, D., Schiffner, S. (eds.) APF 2015. LNCS, vol. 9484, pp. 199–212. Springer, Cham (2016). https://doi.org/10.1007/978-3-319-31456-3_12
5. von Lewinski, K., Pohl, D.: Kommunikation von Datenschutz - Recht und (gute) Praxis. Stiftung Datenschutz, June 2017
6. Chowdhury, O., et al.: Privacy promises that can be kept: a policy analysis method with application to the HIPAA privacy rule. In: Proceedings of the 18th ACM Symposium on Access Control Models and Technologies, SACMAT 2013, pp. 3–14. ACM, New York (2013)
7. Shmueli, E., Tassa, T.: Privacy by diversity in sequential releases of databases. Inf. Sci. **298**, 344–372 (2015)
8. Sweeney, L.: k-anonymity: a model for protecting privacy. Int. J. Uncertain. Fuzziness Knowl. Based Syst. **10**(05), 557–570 (2002)
9. Machanavajjhala, A., Kifer, D., Gehrke, J., Venkitasubramaniam, M.: L-diversity: privacy beyond k-anonymity. ACM Trans. Knowl. Discov. Data **1**(1) (2007). https://doi.org/10.1145/1217299.1217302. Article no. 3
10. Li, N., Li, T., Venkatasubramanian, S.: t-closeness: Privacy beyond k-anonymity and l-diversity. In: Proceedings IEEE 23rd International Conference on Data Engineering, pp. 106–115, April 2007
11. Bertino, E., Lin, D., Jiang, W.: A survey of quantification of privacy preserving data mining algorithms. In: Aggarwal, C.C., Yu, P.S. (eds.) Privacy-Preserving Data Mining. Advances in Database Systems, vol. 34, pp. 183–205. Springer, Boston (2008). https://doi.org/10.1007/978-0-387-70992-5_8
12. Fabian, B., Göthling, T.: Privacy-preserving data warehousing. Int. J. Bus. Intell. Data Min. **10**(4), 297–336 (2015)

13. Xiao, X., Tao, Y.: Personalized privacy preservation. In: Proceedings of the 2006 ACM SIGMOD International Conference on Management of Data, SIGMOD 2006, pp. 229–240. ACM, New York (2006)
14. Kumaraguru, P., Cranor, L., Lobo, J., Calo, S.: A survey of privacy policy languages. In: Workshop on Usable IT Security Management (USM 2007) at Symposium On Usable Privacy and Security 2007 (2007)
15. Kasem-Madani, S., Meier, M.: Security and privacy policy languages: a survey, categorization and gap identification. CoRR, abs/1512.00201 (2015)
16. Hada, S., Kudo, M.: XML access control language: provisional authorization for XML documents, October 2000
17. Damianou, N., Dulay, N., Lupu, E., Sloman, M.: The ponder policy specification language. In: Sloman, M., Lupu, E.C., Lobo, J. (eds.) POLICY 2001. LNCS, vol. 1995, pp. 18–38. Springer, Heidelberg (2001). https://doi.org/10.1007/3-540-44569-2_2
18. Kagal, L.: Rei: a policy language for the me-centric project. Technical report, HP Labs (2002)
19. Bauer, L., Ligatti, J., Walker, D.: Composing security policies with polymer. SIGPLAN Not. **40**(6), 305–314 (2005)
20. Becker, M.Y., Fournet, C., Gordon, A.D.: SecPAL: design and semantics of a decentralized authorization language. J. Comput. Secur. **18**(4), 619–665 (2010)
21. Khandelwal, A., Bao, J., Kagal, L., Jacobi, I., Ding, L., Hendler, J.: Analyzing the AIR language: a semantic web (production) rule language. In: Hitzler, P., Lukasiewicz, T. (eds.) RR 2010. LNCS, vol. 6333, pp. 58–72. Springer, Heidelberg (2010). https://doi.org/10.1007/978-3-642-15918-3_6
22. Lockhart, H., Rissanen, E., Parducci, B.: eXtensible access control markup language (XACML) version 3.0. Technical report, OASIS (2013)
23. Aktug, I., Naliuka, K.: ConSpec - a formal language for policy specification. Electron. Notes Theoret. Comput. Sci. **197**(1), 45–58 (2008)
24. Lamanna, D.D., Skene, J., Emmerich, W.: Slang: a language for defining service level agreements. In: Proceedings of the Ninth IEEE Workshop on Future Trends of Distributed Computing Systems, FTDCS 2003, pp. 100–106, May 2003
25. Meland, P.H., Bernsmed, K., Jaatun, M.G., Castejón, H.N., Undheim, A.: Expressing cloud security requirements for SLAs in deontic contract languages for cloud brokers. Int. J. Cloud Comput. **3**(1), 69–93 (2014). PMID: 58831
26. Oberle, D., Barros, A., Kylau, U., Heinzl, S.: A unified description language for human to automated services. Inf. Syst. **38**(1), 155–181 (2013)
27. Cranor, L., et al.: The platform for privacy preferences 1.1 (P3P1.1) specification. Technical report, W3C (2006)
28. Bohrer, K., Holland, B.: Customer profile exchange (CPExchange) specification, Version 1.0, October 2000
29. Cranor, L., Langheinrich, M., Marchiori, M.: A P3P preference exchange language 1.0 (APPEL1.0). Technical report, W3C (2002)
30. Agrawal, R., Kiernan, J., Srikant, R., Xu, Y.: XPref: a preference language for P3P. Comput. Netw. **48**(5), 809–827 (2005). Web Security
31. Biskup, J., Brüggeman, H.H.: The personal model of data: towards a privacy-oriented information system. Comput. Secur. **7**(6), 575–597 (1988)
32. Ashley, P., Hada, S., Karjoth, G., Schunter, M.: E-P3P privacy policies and privacy authorization. In: Proceedings of the 2002 ACM Workshop on Privacy in the Electronic Society, WPES 2002, pp. 103–109. ACM, New York (2002)

33. Ashley, P., Hada, S., Karjoth, G., Powers, C., Schunter, M.: Enterprise privacy authorization language (EPAL 1.2). Technical report, IBM (2003). https://www.zurich.ibm.com/security/enterprise-privacy/epal/Specification/

34. Ardagna, C., et al.: PrimeLife policy language. In: W3C Workshop on Access Control Application Scenarios. W3C (2009)

35. Yang, J., Yessenov, K., Solar-Lezama, A.: A language for automatically enforcing privacy policies. In: Proceedings of the 39th Annual ACM SIGPLAN-SIGACT Symposium on Principles of Programming Languages, POPL 2012, pp. 85–96. ACM, New York (2012)

36. Schulzrinne, H., Tschofenig, H., Cuellar, J.R., Polk, J., Morris, J.B., Thomson, M.: Geolocation policy: a document format for expressing privacy preferences for location information. RFC 6772, January 2013

37. He, X., Machanavajjhala, A., Ding, B.: Blowfish privacy: tuning privacy-utility trade-offs using policies. In: Proceedings of the 2014 ACM SIGMOD International Conference on Management of Data, SIGMOD 2014, pp. 1447–1458. ACM, New York (2014)

38. Turner, K.J., Reiff-Marganiec, S., Blair, L., Campbell, G.A., Wang, F.: APPEL: an adaptable and programmable policy environment and language. Technical report, Computing Science and Mathematics, University of Stirling, April 2014

39. Azraoui, M., Elkhiyaoui, K., Önen, M., Bernsmed, K., De Oliveira, A.S., Sendor, J.: A-PPL: an accountability policy language. In: Garcia-Alfaro, J., et al. (eds.) DPM/QASA/SETOP-2014. LNCS, vol. 8872, pp. 319–326. Springer, Cham (2015). https://doi.org/10.1007/978-3-319-17016-9_21

40. Prasser, F., Kohlmayer, F., Kuhn, K.A.: A benchmark of globally-optimal anonymization methods for biomedical data. In: 2014 IEEE 27th International Symposium on Computer-Based Medical Systems, pp. 66–71, May 2014

41. Fung, B.C.M., Wang, K., Chen, R., Yu, P.S.: Privacy-preserving data publishing: a survey of recent developments. ACM Comput. Surv. **42**(4), 14:1–14:53 (2010)

42. Yu, T., Li, N., Antón, A.I., A formal semantics for P3P. In: Proceedings of the 2004 Workshop on Secure Web Service, SWS 2004, pp. 1–8. ACM, New York (2004)

43. Sandhu, R.S., Coyne, E.J., Feinstein, H.L., Youman, C.E.: Role-based access control models. Computer **29**(2), 38–47 (1996)

Quantifying and Propagating Uncertainty in Automated Linked Data Integration

Klitos Christodoulou[2]([⊠]), Fernando Rene Sanchez Serrano[1],
Alvaro A. A. Fernandes[1], and Norman W. Paton[1]

[1] School of Computer Science, University of Manchester, Oxford Road,
Manchester M13 9PL, UK
{sanchezf,a.fernandes,norman.paton}@cs.manchester.ac.uk
[2] Department of Information Sciences, Neapolis University Pafos,
2 Danais Avenue, Paphos, Cyprus
klitos@nup.ac.cy

Abstract. The Web of Data consists of numerous Linked Data (LD)
sources from many largely independent publishers, giving rise to the
need for data integration at scale. To address data integration at scale,
automation can provide candidate integrations that underpin a pay-as-
you-go approach. However, automated approaches need: (i) to operate
across several data integration steps; (ii) to build on diverse sources of
evidence; and (iii) to contend with uncertainty. This paper describes the
construction of probabilistic models that yield degrees of belief both on
the equivalence of real-world concepts, and on the ability of mapping
expressions to return correct results. The paper shows how such models
can underpin a Bayesian approach to assimilating different forms of evidence: *syntactic* (in the form of similarity scores derived by string-based
matchers), *semantic* (in the form of semantic annotations stemming from
LD vocabularies), and *internal* in the form of fitness values for candidate
mappings. The paper presents an empirical evaluation of the methodology described with respect to equivalence and correctness judgements
made by human experts. Experimental evaluation confirms that the proposed Bayesian methodology is suitable as a generic, principled approach
for quantifying and assimilating different pieces of evidence throughout
the various phases of an automated data integration process.

Keywords: Probabilistic modelling · Bayesian updating
Data integration · Linked Data

1 Introduction

There has been a general trend towards generating large volumes of data, especially with the explosion of social media and other sensory data from smart
devices. The Web is no exception to the accelerating and unprecedented rate
at which digital data is being generated. Because of this explosion, data is now
made available with different characteristics: with different degrees of structure

© Springer-Verlag GmbH Germany, part of Springer Nature 2018
A. Hameurlain and R. Wagner (Eds.): TLDKS XXXVII, LNCS 10940, pp. 81–112, 2018.
https://doi.org/10.1007/978-3-662-57932-9_3

(e.g., structured or unstructured), often semantically annotated (e.g., Linked Data (LD)), typically stored in various distributed data sources[1], often designed independently using different data models, and maintained autonomously by different actors. This makes it imperative to integrate data from various sources with the aim of providing transparent querying facilities to end-users [19]. However, this integration task poses several challenges due to the different types of heterogeneities that are exhibited by the underlying sources [10,12]. For instance, in the case of the Web of Data (WoD), LD sources do not necessarily adhere to any specific, uniform structure and are, thus, considered to be schema-less [5]. This can lead to a great diversity of publication processes, and inevitably means that resources from the same domain may be described in different ways, using different terminologies.

The challenging problem of resolving the different kinds of heterogeneities that data sources exhibit with the aim of providing a single, transparent interface for accessing the data is known as *data integration* [10,12]. A *traditional* data integration system [19] builds on a *mediator*-based architecture where a virtual schema is designed that captures the integration requirements and is presented to the user for querying. In this approach, the *integration* schema is seen as a logical schema since the data still resides in the underlying data sources (as opposed to being materialized, as is typically the case for data warehouses). Typically, for the underlying sources to interoperate, two basic capabilities are required: (i) *matching*, i.e., the ability to quantify the degree of similarity between the source schemas and the integration schema (often by considering their terminologies, and, if available, samples of instance data), the result of which is a set of semantic correspondences (a.k.a. matches); and (ii) *mapping generation*, i.e., the ability to use the set of semantic correspondences in order to derive a set of executable expressions (a.k.a. mappings) that, when evaluated, translate source instance data into instance data that conforms to the integration schema.

Dataspaces are data integration systems that build on a *pay-as-you-go* approach for incremental and gradual improvement of automatically derived speculative integrations [20,30]. In this approach, the manual effort required to set up a traditional data integration system is replaced with automatic techniques that aim to generate a sufficiently useful initial integration with minimum human effort [11,14]. Over time, as the system is continuously queried, users are stimulated to provide feedback (e.g., on query results) that, once assimilated, lead to a gradual improvement in the quality of the integration [2]. More specifically, with a view to providing *best-effort* querying capabilities, dataspaces are envisioned to have a *life-cycle* (depicted in Fig. 1) comprising the following phases: (i) *bootstrapping*, where algorithmic techniques are used to automatically derive an initial integration by postulating the required semantic correspondences and using them to derive mappings between the source schemas and the integra-

---

[1] One well-known example portal is the so-called Linked Open Data (LOD) cloud, at https://lod-cloud.net/.

tion schema[2]; (ii) *usage*, where best-effort querying services are provided to answer user requests over the speculative integration, and explicit or implicit feedback [24] is collected to inform the incremental improvement of the integration; and finally (iii) *improvement*, where the feedback that has been collected during *usage* is assimilated in order to improve the initial integration, e.g., by filtering erroneously-derived semantic correspondences and regenerating the mappings previously derived from them.

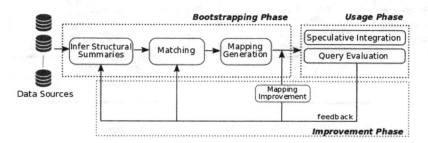

Fig. 1. *Dataspace* life-cycle phases.

Because dataspaces depend on automation, and because automation can only generate inherently uncertain outcomes, it is imperative to quantify and propagate uncertainty throughout the dataspace life-cycle [18,31]. Broadly speaking, it is not obvious how the inherent uncertainty arising during the various phases of a dataspace system can be quantified and then reasoned with in a principled manner. Motivated by this challenge, and taking into account the different types of uncertainty that must be quantified and propagated across the phases of the dataspace life-cycle, this paper contributes a methodology for quantifying uncertainty (founded on the construction of empirical probabilistic models based on *kernel estimators*) and for reasoning with different kinds of evidence that emerge during the *boostrapping phase* of a dataspace system using Bayesian techniques for assimilating: (a) *syntactic evidence*, in the form of similarity scores generated by string-based matchers, (b) *semantic evidence*, in the form of semantic annotations such as subclass-of and equivalent relations that have been asserted in, or inferred from, LD ontologies, and (c) *internal evidence*, in the form of *mapping fitness values*, produced during mapping generation.

1.1   Motivating Example: Uncertainty in Dataspaces

Our motivating example comes from the music domain. We assume the inferred schemas for the *Jamendo* LD source[3], denoted by s_1, and the *Magnatune* LD

---

[2] For schema-less sources (e.g., Linked Data sources) schema extraction techniques can be used to infer schemas (e.g., [5]).

[3] https://www.jamendo.com/.

source[4], denoted by s_2, depicted in simplified ER notation in Fig. 2 (a) and (b), respectively. The goal is integrate these to give rise to an integrated schema, denoted by s_{int}.

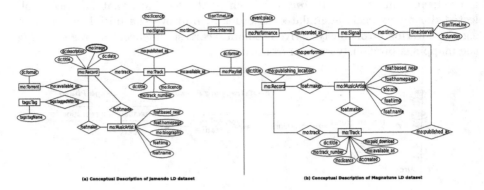

Fig. 2. Inferred schemas from LD sources.

For the identification of these candidate semantic matches, several approaches have been proposed especially by the literatures on *schema matching* [27] in the database area, and on *ontology alignment* [32] in the knowledge representation area. Figure 3 shows a subset of semantic correspondences (i.e., matches) that might have been discovered across our example schemas using string-based matching techniques (e.g., n-gram).

$mt_1 : \langle s_1.Record,\ s_{int}.Record,\ 1.0 \rangle$
$mt_2 : \langle s_1.Record.title,\ s_{int}.Record.track\_title,\ 0.54 \rangle$
$mt_3 : \langle s_1.Tag.tagName,\ s_{int}.MusicArtist.name,\ 0.45 \rangle$
$mt_4 : \langle s_2.Performance.recorded\_as,\ s_{int}.Record,\ 0.6 \rangle$

Fig. 3. Example schema matching results.

Figure 4 exemplifies different kinds of semantic evidence regarding our example schemas. In this figure, solid arrows denote annotations (e.g., rdfs:sub ClassOf) either internal, pointing to constructs in the same LD vocabulary, or external, pointing to constructs in some other LD vocabulary; dashed arrows denote *equivalence* annotations that define entities; and dotted lines show examples of one-to-one semantic correspondences where confidence is measured as a d.o.b. As this example shows, semantic relationships may exist in addition to syntactic ones, e.g., *mo:MusicGroup* is also subsumed by *foaf:Group* and not simply named in a syntactically similar way to the latter. Section 3.2 presents a methodology for quantifying semantic evidence that is founded on the construction of probabilistic models that can be used to inform a Bayesian approach for making judgements on the equivalence of constructs.

---

[4] http://magnatune.com/.

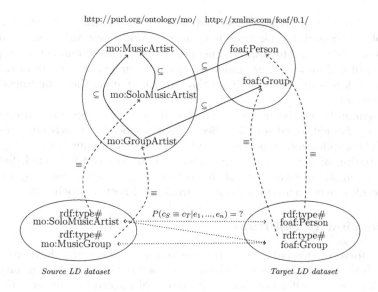

Fig. 4. Different kinds of semantic evidence.

Table 1 shows examples of such internal evidence *viz.*, where the fitness values and corresponding mapping correctness score (as explained in Sect. 2) are assumed to have been returned by the mapping generation process. Some examples of mapping queries are provided in Table 2.

Table 1. Example of internal evidence from the mapping generation phase.

map_id	Target	Source	Fitness	Mapping correctness
$m_1 : \langle\rangle$	Solomusicartist	Musicartist	0.845	0.86
$m_2 : \langle\rangle$	Track	Track	0.256	0.33
$m_3 : \langle\rangle$	Musicgroup	Musicartist	0.92	0.86
$m_4 : \langle\rangle$	Lyrics	Performance	0.0048	0

1.2 Summary of Contributions

This paper describes a probabilistic approach for combining different types of evidence so as to annotate integration constructs with *d.o.b.s* on semantic equivalence and on mapping quality. This paper contributes the following: (a) a methodology that uses kernel density estimation for deriving likelihoods from similarity scores computed by string-based matchers; (b) a methodology for deriving likelihoods from semantic relations (e.g., rdfs: subClassOf, owl:equivalentClass) that are retrieved by dereferencing URIs in LD ontologies; (c) a methodology for aggregating evidence of conceptual equivalence of constructs from both string-based matchers and semantic annotations; (d) a methodology for deriving likelihoods from mapping fitness values and mapping correctness scores using bivariate kernel density estimation; and (e) an empirical evaluation of our approach grounded on the judgements of experts in response to the same kinds of evidence. Note that, in this paper, the experiments only use LD datasets.

The remainder of the paper is structured as follows. Section 2 presents an overview of the developed solution. Section 3 describes the contributed methodologies. The application of bayesian updating, as a technique for the incremental assimilation of data integration evidence, is introduced in Sect. 4. Section 5 presents an empirical evaluation of the methodology complemented by a discussion of results. Section 6 reviews related work, and Sect. 7 concludes.

2   Overview of Solution

The main focus of this paper is on the *bootstrapping phase* of a data integration system. More specifically, the techniques discussed in this section focus on opportunities for the quantification and assimilation of uncertainty using a *Bayesian* approach to assimilate different forms of evidence. Figure 5 stands in contrast with Fig. 1 and indicates the different types of evidence that inform different bootstrapping stages in our approach. Our techniques have been implemented as extensions to the *DSToolkit* [13] dataspace management system, which brings together a variety of algorithmic techniques providing support for the dataspace life-cycle.

Deriving d.o.b.s on Matches. Assuming that declared (or else inferred) conceptual descriptions (e.g., schemas) for the sources and target integrated artefact are available, the basic notion underpinning the bootstrapping process is that of semantic correspondences. Given a conceptual description of a *source* and a *target* LD dataset, denoted by S and T, respectively, a *semantic correspondence* is a triple $\langle c_S, c_T, P(c_S \equiv c_T | E) \rangle$, where $c_S \in S$ and $c_T \in T$ are constructs (e.g., classes, or entity types) from (the schema of) the datasets, and $P(c_S \equiv c_T | E)$ is the conditional probability representing the d.o.b. in the equivalence ($\equiv$) of the constructs given the pieces of evidence $(e_1, \ldots, e_n) \in E$. Such semantic correspondences, therefore, quantify (as d.o.b.s) the uncertainty resulting from automated matching techniques that yield syntactic evidence in the form of similarity scores but also taking into account, when available, semantic annotations from ontologies (as exemplified in Fig. 4).

Deriving d.o.b.s on Mappings. Given a set of *semantic correspondences*, the mapping generation process derives a set of mappings M using (in the case of DSToolkit) an evolutionary search strategy that assigns a *fitness value* to each mapping in the solution set. A mapping $m \in M$ denotes that one or more schema constructs from S that can be used to populate one or more constructs from T. The two sets of schema constructs that are related in this way by a mapping are henceforth referred to as *entity sets* and notated as $\langle ES^S, ES^T \rangle$, where $ES^S \in S$ and $ES^T \in T$.

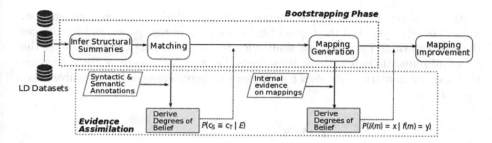

Fig. 5. Uncertainty propagation and evidence assimilation.

Table 2. Example generated mappings.

Mapping	Fitness Value
$m_1 : \langle s_{int}.Record,$ **SELECT** R.title as title, R.maker as maker, NULL as description, NULL as date_created, T.title as track_title, T.paid_download as paid_download **FROM** s_2.Record R, s_2.Track T **WHERE** R.track = T.title$\rangle$	0.42459
$m_2 : \langle s_{int}.SoloArtist,$ **SELECT** M.name as name, M.img as img, NULL as biography, M.homepage as homepage, M.based_near as based_near **FROM** s_2.MusicArtist M$\rangle$	0.84560

As a result of the search technique used in the mapping generation process, a mapping fitness value y is a measure of the strength of the internal evidence that a set of schema constructs in a source entity set ES^S is semantically related to a set of schema constructs in the target entity set ES^T. In this context, $P(m \mid f(m) = y)$ is the conditional probability representing the d.o.b. that an attribute value in a tuple returned by the mapping m is likely to be correct, given that the fitness value of m, is y i.e., $f(m) = y$. Such probabilities, therefore, quantify (as d.o.b.s) the uncertainty resulting from the automated mapping generation technique used, i.e., one that yields mapping fitness values. Table 2 shows some mappings generated between a *target* schema, denoted by s_{int} and *source* schemas, denoted by s_1, and s_2, resp., along with their associated *fitness value* scores.

Types of Evidence. As indicated above, our approach makes use of three distinct types of evidence: (a) *syntactic evidence*, in the form of strings that are local-names of resource URIs; (b) *semantic evidence*, such as structural relations between entities, either internal to a vocabulary or across different LD vocabularies (e.g., relationships such as subclass of and equivalence); and (c) *internal* *evidence*, in the form of fitness values computed during mapping generation. Table 3 briefly describes the types of evidence used in this paper and introduces the abbreviations by which we shall refer to them. In particular, if TE is the set of all semantic annotations, its subsets EE and NE comprise the assertions that can be construed as *direct* evidence of equivalence and non-equivalence, respectively.

Table 3. Types of evidence.

Type		ID	Description	Evidence rule
Syntactic evidence (LE)	-	SLN	similar-local-name	$string\ similarity(c_T, c_S)$
Semantic evidence (TE)		SU	same-URI	$string\ equality(URI_S, URI_T)$
	-	SB	subsumed-by	$c_S \sqsubseteq c_T$
	EE	SA	same-as	owl:sameAs(c_S, c_T)
		EC	equivalent-class	owl:equivalentClass(c_S, c_T)
		EM	exact-match	skos:exactMatch(c_S, c_T)
	NE	DF	different-from	owl:differentFrom(c_S, c_T)
		DW	disjoint-with	owl:disjointWith(c_S, c_T)
Mapping generation evidence	-	MGE	mapping fitness value	$fitness\ value(ES^S, ES^T)$

Collecting Evidence. To collect syntactic evidence (represented by the set LE), given two sources, our approach extracts local names from the URIs of every pair of constructs $\langle c_s, c_t \rangle$ and then derives their pairwise string-based degree of similarity. Two string-based metrics are used in our experiments, viz., *edit-distance* (denoted by ed) and *n-gram* (denoted by ng) [32]. Section 3.1 explains in detail how probability distributions can be constructed for each matcher. To collect semantic evidence, our approach dereferences URIs to obtain access to annotations from the vocabularies that define the resource. For example, the subsumption relation $c_S \sqsubseteq c_T$ is taken as semantic evidence. Section 3.2 explains in detail how to construct probability distributions for each kind of semantic evidence published in RDFS/OWL vocabularies. To collect evidence on mapping generation, we extract from the set of mappings generated by a mapping generation algorithm their fitness value, and, therefore, we assume that the search procedure underpinning the algorithm aims to maximize an objective function founded on such fitness values [8]. Section 3.3 explains in detail how to construct probability distributions for mapping fitness values. Later in our methodology, the probability distributions thus constructed are used to denote the likelihood of evidence term in Bayes's formula.

We use a Bayesian approach to evidence assimilation, i.e., given a degree of uncertainty expressed as a d.o.b., once new evidence is observed, we use Bayes's formula to update that d.o.b. (referred to as the *prior*) into a new d.o.b. (referred to as the *posterior*) that reflects the new evidence. Applying Bayes's formula in this way requires us to quantify the uncertainty of the evidence (referred to as the *likelihood*). This means that in order to assimilate different kinds of evidence, preliminary work is needed to enable the computation of the likelihoods in applications of Bayes's formula, i.e., the otherwise unknown term required for the calculation of a posterior d.o.b. from a prior d.o.b. This requirement holds for the equivalence of constructs, as captured by the posterior $P(c_S \equiv c_T | E)$ when the evidence is syntactic (as described in Sect. 3.1) and when the evidence is semantic (as described in Sect. 3.2). Similarly, preliminary work is needed for deriving a d.o.b. on mapping correctness, as captured by the posterior $P(m \mid f(m) = y)$, where $f(m) = y$ is internal evidence from mapping generation in which we relate the notion of mapping correctness to the fraction of correct attribute values in a mapping extent (as described in Sect. 3.3).

The idea behind *Bayesian updating* [34] is that once the posterior (e.g., $P(c_S \equiv c_T | E)$) is computed for some evidence $e_1 \in E$, a new piece of evidence $e_2 \in E$ allows us to compute the impact of e_2 (i.e., measure how the d.o.b. is changed in light of e_2) by taking the previously computed posterior as the new prior.

3   Constructing Likelihoods for Evidence

We now provide a detailed account of a principled methodology for constructing probability distributions from relevant evidence with a view towards enabling a Bayesian approach to quantifying and propagating uncertainty across the dataspace life-cycle.

3.1   Deriving Likelihoods for Similarity Scores

We call *syntactic evidence* the likelihoods derived from *similarity scores* produced by string-based matchers. We study the behaviour of each matcher (in our case ed and ng) used to derive similarity scores.

To derive probability density functions (PDFs) for syntactic evidence, we proceeded as follows:

1. From the datasets made available by the Ontology Alignment Evaluation Initiative (OAEI)[5], we observed the available ground truth on whether a pair of local-names, denoted by (n, n'), aligns.
2. We assumed the existence of a continuous random variable, X, in the bounded domain $[0, 1]$, for the similarity scores returned by each matcher μ, where $\mu \in \{ed, ng\}$. Our objective was to model the behaviour of each matcher in terms of a PDF $f(x)$ over the similarity scores it returns, which we refer to as observations in what follows.

---

[5] http://oaei.ontologymatching.org.

3. To empirically approximate $f(x)$ for each matcher, we proceeded as follows:
   (a) We ran each matcher μ independently over the set of all local-name pairs (n, n') obtained from (1).
   (b) For each pair of local-names, we observed the independent similarity scores returned by the matcher when (n, n') agrees with the ground truth. These are the set of observations $(x_1, \ldots, x_i)$ from which we estimate $f(x)$ for the equivalent case.
4. The observations $x_1, \ldots, x_i$ obtained were used as inputs to the non-parametric technique known as kernel density estimation (KDE) (using a Gaussian kernel[6]) [4] whose output is an approximation $\hat{f}(x)$ for both ed and ng and for both the equivalent and non-equivalent cases.

We interpret the outcome of applying such a PDF to syntactic evidence as the likelihood of that evidence. More formally, and as an example, $PDF \underset{ed}{\equiv} (\text{ed}(n, n')) = P(\text{ed}(n, n')|c_S \equiv c_T)$, i.e., given a pair of local-names (n, n') the PDF for the ed matcher in the equivalent case $PDF \underset{ed}{\equiv}$ yields the likelihood that the similarity score $\text{ed}(n, n')$ expresses the equivalence of the pair of concepts (c_S, c_T) that (n, n'), resp., denote. Correspondingly, for the non-equivalent case, and for ng in both the equivalent and non-equivalent cases.

The PDFs derived by the steps described above are shown in Fig. 6(a) and (b) for ed and in Fig. 6(c) and (d) for ng. The same procedure can be used to study the behaviour of any matcher that returns similarity scores in the interval $[0, 1]$. Note that the PDFs obtained by the method above are *derive-once, apply-many* constructs. Assuming that the samples used in the estimation of the PDFs remain representative, and given that the behaviour of matchers ed and ng is fixed and deterministic, the PDFs need not be recomputed.

3.2   Deriving Likelihoods for Semantic Evidence

We call *semantic evidence* the likelihoods derived from *semantic annotations* obtained from the WoD. We first retrieved the semantic annotations summarised in Table 3. The set TE is the set of all such evidence, $TE = SU \cup SB \cup SA \cup EC \cup EM \cup DF \cup DW$. We formed the subsets $EE \subset TE = SA \cup EC \cup EM$ and $NE \subset TE = DF \cup DW$ comprising assertions that can be construed as *direct* evidence of equivalence and non-equivalence, respectively.

To derive a PDF for semantic evidence, we proceeded as follows:

1. We assumed the existence of a Boolean random variable, for each type of semantic evidence in Table 3, with domain $\{true, false\}$.
2. Using the vocabularies available in the Linked Open Vocabularies (LOV)[7] collection, we collected and counted pairs of classes and properties that share

---

[6] A Gaussian kernel was used due to its mathematical convenience. Note that any other kernel can be applied. Of course, the shape of the distribution may differ depending on the kernel characteristics.

[7] http://lov.okfn.org/dataset/lov/.

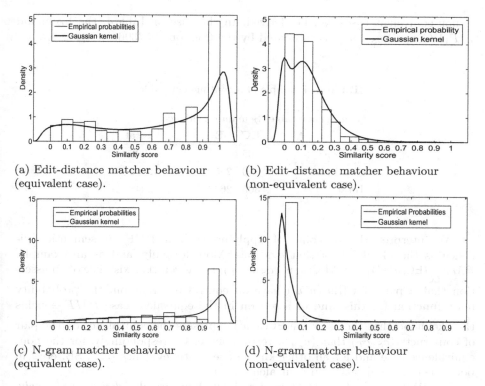

(a) Edit-distance matcher behaviour (equivalent case).

(b) Edit-distance matcher behaviour (non-equivalent case).

(c) N-gram matcher behaviour (equivalent case).

(d) N-gram matcher behaviour (non-equivalent case).

Fig. 6. Illustration of probability distributions for each matcher over $[0, 1]$.

direct or indirect assertions of equivalence or non-equivalence for all the assertions in TE and NE using SPARQL queries. For example, with respect to equivalence based on OWL and RDFS class annotations:

```
SELECT DISTINCT ?elem1 ?elem2
WHERE {
    {?elem1 a rdfs:Class .} UNION {?elem1 a owl:Class .}
    ?elem1 ?p ?elem2 .
    FILTER (?p = owl:equivalentClass && !isBlank(?elem2)) }
```

3. From the set of pairs derived by the assertions in TE and NE, we counted assertions that can be construed as *evidence* of equivalence or non-equivalence for each pair, grouping such counts by kind of assertion (e.g., subsumed-by, etc.)
4. We used the sets of counts obtained in the previous step to build contingency tables (as exemplified by Table 4) from which the probability mass functions (PMFs) for each kind of semantic evidence for both the equivalence and

non-equivalent cases can be derived. In the case of Table 4, the likelihood $P(\mathbf{EC}(n, n')|c_S \equiv c_T)$ is estimated by the fraction $305/396$.

Table 4. Example of a contingency table.

Contingency table	Semantic evidence		
	EC	**¬EC**	Total
$c_S \equiv c_T$	305	91	396
$c_S \not\equiv c_T$	0	2552	2552
Total	305	2643	2948

We interpret the outcome of applying such a PMF to semantic evidence as the likelihood of that evidence. More formally, and as an example, $PMF_{\underset{EC}{\equiv}}(\mathsf{EC}(u, u')) = P(\mathsf{EC}(u, u')|c_S \equiv c_T)$, i.e., given the existence of an assertion that a pair of URIs (u, u') have an equivalence relation, the probability mass function for this kind of assertion in the equivalent case $PMF_{\underset{EC}{\equiv}}$ yields the likelihood that the assertion $\mathsf{EC}(u, u')$ expresses the equivalence on the pair of constructs (c_S, c_T) that (u, u'), resp., denote. Correspondingly, for the non-equivalence case and for all other kinds of semantic evidence (e.g., SB, etc.) in both the equivalent and non-equivalent cases.

The PMFs derived by the steps described above are also *derive-once, apply-many* constructs, but since the vocabulary collection from which we draw our sample is dynamic, it is wise to be conservative and view them as *derive-seldom, apply-often*.

3.3 Deriving Likelihoods for Internal Evidence

We call *internal evidence* the likelihoods derived from *mapping fitness values* returned by the mapping generation process.

Note that in quantifying the uncertainty in respect of matching outcomes, the hypothesis of equivalence can be modelled as a Boolean random variable. However, in the case of mapping outcomes, this binary classification is undesirable. In the case that we adopt a binary setting with two possible outcomes, viz., correct or incorrect, a correct mapping would be one that produces exactly the same extent as the ground truth, any other mapping would be deemed incorrect. However, a mapping may still be useful even if it fails to produce a completely correct result. In practice, requiring mappings to be correct in this most stringent sense may lead to few correct mappings whilst ruling out many useful mappings. Therefore, for mapping outcomes, rather than expecting a pair of constructs to be either equivalent or not, we are interested in the *degree of correctness of a mapping*, and, therefore, we start by associating a *mapping correctness score* to a mapping.

More formally, we denote by $[\![m]\!]$ the *extent* of m, i.e., the result of evaluating m over some instance and introduce a measure $\delta(m)$ that assigns a degree of correctness m as the fraction of correct attribute values in $[\![m]\!]$, where $\delta(m) \in [0,1]$. This measure can be computed for a mapping m and the ground truth GT (taken as an instance) based on the number of identical attribute values between $[\![m]\!]$ and GT as follows:

$$S(m) = \sum_{i=1}^{|GT|} max_{j=1...|[\![m]\!]|} \left(t_{sim}(t_{GT_i}, t_{m_j})\right) \tag{1}$$

$$S'(m) = \sum_{j=1}^{|[\![m]\!]|} max_{i=1...|GT|} \left(t_{sim}(t_{GT_i}, t_{m_j})\right) \tag{2}$$

$$\delta(m) = \frac{S(m) + S'(m)}{|GT| + |[\![m]\!]|} \tag{3}$$

where $[\![m]\!]$ is the set of tuples resulting from the evaluation of m over GT, and $t_{sim}()$ is a function that computes the similarity between two tuples as the ratio of identical attribute-aligned values as follows:

$$t_{sim}(t_{GT}, t_m) = \frac{|\{a \in t_{GT}|t_{GT}(a) = t_m(b), aligned(a,b)\}|}{arity(GT)} \tag{4}$$

where a and b are attributes belonging to GT and $[\![m]\!]$, resp., $t_{GT}(a)$ is the value of the attribute a, $t_m(b)$ is the value of the attribute b, and $aligned(a,b)$ is true iff a and b are considered to be a match (i.e., there is a postulated conceptual equivalence between a and b). Intuitively, $S(m)$ estimates how similar the tuples in GT are to the tuples in m, and $S'(m)$ estimates how similar the tuples in m are to those in the GT, whereas $\delta(m)$ combines these estimates.

With the goal of deriving likelihoods from internal evidence in the form of mapping fitness values, we then correlate the latter with the corresponding mapping correctness scores.

We must study the distribution of mapping fitness values for a comprehensive set of mappings showing different fractions of correct attribute values in their extent. We used a comprehensive set of paired observations obtained from a diverse set of integration scenarios. Using a representative sample of observations as input to KDE leads to a better estimate of the unknown distribution [33]. To collect as many observations as possible, we exposed the integration tool (i.e., DSToolkit) to as many types of heterogeneities as are likely to be found in real-world integration scenarios.

We used MatchBench [9] to systematically inject into an initial schema various heterogeneities (at the entity, and the attribute levels) between two sets of schema constructs under the classification proposed by Kim *et al.* [17]. Examples of schematic heterogeneities include missing attributes, inconsistent naming, as well as horizontal and vertical partitionings.

In more detail, in order to derive probability density functions (PDFs) for internal evidence, we proceeded as follows:

1. Given a pair of initial schemas (S, T), we injected a set of systematic hetero-geneities into the initial schemas as described in [9], where for each hetero-geneity introduced, so as to derive, using MatchBench, a new pair of schemas (S', T') that reflects the changes intended for that scenario.
2. For every new pair of schemas (S', T'), and a set of matches between S' and T', we derived, using DSToolkit, a set of mappings M' between S' and T'.
3. For each mapping $m_i \in M'$, we observed its fitness value $f(m_i) = y$ and computed its degree of correctness $\delta(m_i) = x$, based on the extent produced by the mapping m_i and the corresponding ground truth GT (constructed by hand), giving rise to a pair of measures (x_i, y_i) which we refer to as observations in what follows.
4. We assumed the existence of a continuous random variable $X \in [0, 1]$ for the correctness score of a generated mapping.
5. We assumed the existence of a continuous random variable $Y \in [0, 1]$ for the fitness value associated with a generated mapping.
6. The observations $x_1, \ldots, x_i$, and $y_1, \ldots, y_i$, $i = |M|$ obtained as described above were used as inputs to a bivariate KDE (using a Gaussian kernel) whose output is an approximation of the PDF of the two continuous variables, $\hat{f}(x, y)$.

We interpret the outcome of applying such a PDF to this internal evidence as the likelihood of that evidence. The PDF yields the probability of observing a mapping fitness value y given that a mapping has a correctness score x. More formally, this is expressed as $P(f(m) = y \mid \delta(m) = x)$. As with the previous cases, the obtained PDF is a *derive-once, apply-many* construct. Assuming that the sample of mappings used for training remains representative, and highly correlates mapping correctness scores with mapping fitness values, the PDF need not be recomputed. As is the case with semantic evidence, a certain degree of domain dependency suggests it is wise to consider the process one whose type is *derive-seldom, apply-often*. Figure 7 depicts the resulting bivariate PDF.

4   Assimilating Evidence Using Bayesian Updating

The purpose of deriving likelihood models as described in Sect. 3 is to enable the evidence to be combined in a systematic way using Bayesian updating. The procedure for doing so is now described, but the benefits of the procedure are only discussed in Sect. 5.

We denote by S and T the structural summaries (an ontology or a structural summary derived by an approach like [5]) that describe, resp., the structure of a *source* and a *target* LD source (over which we wish to discover semantic correspondences) and that are used to derive a set of mappings between S and T. Firstly, we show how to assimilate syntactic and semantic evidence to postulate a d.o.b. on the equivalence of two constructs. Then, we elaborate on how the Bayesian updating methodology can be used to update the derived posterior in the light of additional evidence that emerge from the mapping generation phase, and thereby postulate a d.o.b. on mapping correctness.

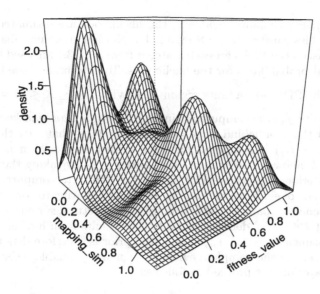

Fig. 7. Bivariate PDF showing the correlation of mapping correctness score with the mapping fitness values.

Assimilating Syntactic and Semantic Evidence on Matches. Given a pair of constructs $c_S \in S$ and $c_T \in T$, our objective is to derive a d.o.b. on the postulated equivalence of a pair of constructs (denoted by H), given pieces of evidence $e_1, \ldots, e_n \in E$. To reason over our hypothesis, we model it as a conditional probability $P(H|E)$ and apply Bayes's theorem to make judgements on the equivalence of two constructs. The classical form of Bayes's theorem[8] is:

$$P(H|E) = \frac{P(E|H)\,P(H)}{P(E)}. \tag{5}$$

To formulate the hypothesis for the matches, we assume a *Boolean hypothesis* to postulate equivalence of constructs. In this case, the hypothesis can take one of two states: $P(H) = \{P(c_S \equiv c_T), P(c_S \not\equiv c_T)\}$. The prior probability, i.e., $P(H) = P(c_S \equiv c_T)$, is the d.o.b. in the absence of any other piece of evidence (we assume a uniform distribution). Thus, since $N = 2$, i.e., there are two possible outcomes our hypothesis can take, the prior probability that one of the outcomes is observed is $1/N$. The probability of the evidence, $P(E)$, can be expressed using the law of total probability [23], i.e., $P(E) = P(E|c_S \equiv c_T)\,P(c_S \equiv c_T) + P(E|c_S \not\equiv c_T)\,P(c_S \not\equiv c_T)$. To use Bayes's theorem for deriving a d.o.b. on the hypothesis given the available evidence, it is essential to estimate the likelihoods for each type of evidence, i.e., $P(E|c_S \equiv c_T)$ and

---

[8] Informally, the d.o.b., in the hypothesis given the evidence (the so-called posterior d.o.b.) is equal to the ratio between the product of the d.o.b. in the evidence given the hypothesis (which we call likelihood in Sect. 3) and the d.o.b. in the hypothesis (the so-called prior d.o.b.) divided by the d.o.b. in the evidence.

$P(E|c_S \not\equiv c_T)$. For semantic evidence, the likelihoods are estimated from the contingency tables constructed in Sect. 3.2. For continuous values, like similarity scores, the constructed PDFs for each matcher from Sect. 3.1 are used to estimate the conditional probabilities for the likelihoods. To determine these likelihoods, we integrate the PDF over a finite region $[a,\, b]$, viz., $P(a \le X \le b) = \int_a^b f(x)\, dx$, where the density $f(x)$ is computed using KDE with a *Gaussian* kernel.

Recall that the idea behind *Bayesian updating* [34] is that once the posterior (e.g., $P(c_S \equiv c_T|E)$) is computed for some evidence $e_1 \in E$, a new piece of evidence $e_2 \in E$ allows us to compute the impact of e_2 by taking the previously computed posterior as the new prior. Given the ability to compute likelihoods for different kinds of evidence, we can use Bayesian updating to compute a d.o.b. on the equivalence of (pairs of constructs in) two structural summaries S and T. To see this, let $P^{(e_1,\ldots,e_n')}$ denote the d.o.b. that results from having assimilated the evidence sequence $(e_1, \ldots, e_n)$. The initial prior is therefore denoted by $P^{()}$, and if $(e_1, \ldots, e_n)$ is the complete evidence sequence available, then $P^{(e_1,\ldots,e_n')}$ is the final posterior. We proceed as follows:

i. We set the initial prior according to the principle of indifference between the hypothesis that $P(c_S \equiv c_T)$ and its negation, so $P^{()} = 0.5$.

ii. We collect the local-name pairs from the structural summaries S and T.

iii. We run ed on the local-name pairs and, using the probability distributions derived using the methodology described above (Sect. 3.1), compute the likelihoods for each pair and use Bayes's rule to calculate the initial posterior $P^{(ed)}$.

iv. We run ng on the local-name pairs and, using the probability distributions derived using the methodology described above (Sect. 3.1), compute the likelihoods for each pair and use Bayes's rule to calculate the next posterior $P^{(ed,ng)}$. Note that this is the d.o.b. given the syntactic evidence alone, which we denote more generally by $P^{(syn)}$.

v. To get access to semantic annotations that span a variety of LD ontologies, we dereference every URI in S and T to collect the available semantic annotations e.g., $SB(c_S \subseteq c_T)$.

vi. Using the methodology described above (Sect. 3.2), we compute, one at a time, the likelihoods for the available semantic evidence, each time using Bayes's rule to calculate the next posterior (e.g., $P^{(ed,ng,SB,\ldots)}$), so that once all the available semantic evidence is assimilated, the final posterior, which we denote more generally by $P^{(syn,sem)}$, is the d.o.b. on $c_S \equiv c_T$, where, $c_S \in S \wedge c_T \in T$.

Before carrying out the empirical evaluation of this approach using syntactic and semantic evidence described in Sect. 5, we studied analytically, using Bayes's theorem, the effect of each piece of evidence independently. Given a series of initial prior probabilities in the range of $[0, 1]$ and the evidence likelihoods (see Sect. 3) we computed the posterior probabilities given each piece of evidence. Figure 8(a) and (b) show how the posteriors $P(c_s \equiv c_t|ed(c_s, c_t) = s)$, and, $P(c_s \equiv c_t | ng(c_s, c_t) = s)$, resp., are updated when the available evidence is similarity scores computed by the string-based matchers ed and ng. As an example, consider Fig. 8(a) and assume that we are given a prior probability of $x = 0.5$ and a similarity score that is $y < 0.5$, ed will cause the updated posterior probability to fall relatively more. In this case, if the similarity score is $y = 0.2$, the posterior probability drops to $z = 0.2$. In the case of ng, using identical values as previously, the posterior probability drops to $z = 0.36$, which means that ng causes a smaller decrease in the posterior than the ed does. In a similar fashion, the independent behaviours of different kinds of semantic evidence have been studied. For example, Fig. 8(c) shows how the posterior is updated when there is direct evidence that a pair of classes stand in a subsumption relationship (i.e., SB). A subsumption relation may indicate that the constructs are more likely to be related than to be disjoint and a low initial prior is therefore increased into a larger posterior. Similarly, Fig. 8(d) shows how the posterior is affected when a pair of constructs stand in an equivalence relation (i.e., EC). This is considered enough evidence to significantly increase a low prior to close to 1; meaning that constructs are much more probably equivalent than if that evidence had not been available.

Having observed how different posterior probabilities are updated in the presence of individual pieces of evidence, in Sect. 5 we empirically assess whether the incorporation of semantic evidence from LD ontologies can improve on judgements on the equivalence of constructs obtained through syntactic matching alone.

Assimilating Evidence on Mappings. Similarly to the matching case, we use the Bayesian updating methodology to revise a previously computed posterior with a d.o.b. on mapping correctness in the light of evidence in the form of fitness values. For this purpose, we postulate our hypothesis as a degree of correctness of a mapping m, denoted as the mapping correctness score $\delta(m) = x$. Therefore the posterior d.o.b. can be expressed using Bayes's theorem:

$$P(\delta(m) = x \mid f(m) = y) = \frac{P(f(m) = y \mid \delta(m) = x)P(\delta(m) = x)}{P(f(m) = y)} \qquad (6)$$

where $P(\delta(m) = x)$ is the prior probability that a mapping m has a degree of correctness x (drawn from a continuous uniform distribution, $U(0, 1)$), and $P(f(m) = y \mid \delta(m) = x)$ is the likelihood of observing a mapping fitness value y for a mapping m, given that m has a degree of correctness x. We use the constructed PDF described in Sect. 3.3 to compute the conditional probability. More specifically, and assuming that $f(m)$ and $\delta(m)$ are two jointly continuous random variables, described in terms of the derived PDF, the likelihood $P(f(m) = y \mid \delta(m) = x)$, can be computed using the definition of conditional probability as follows:

$$P(f(m) = y \mid \delta(m) = x) = \frac{P(f(m) = y \cap \delta(m) = x)}{P(\delta(m) = x)} \qquad (7)$$

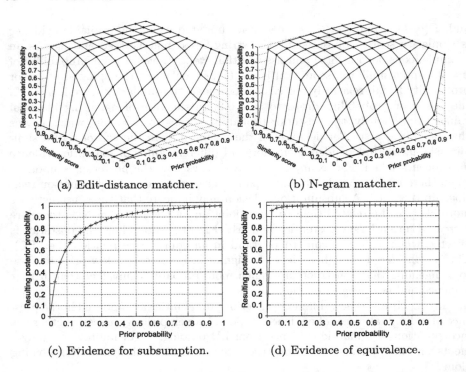

(a) Edit-distance matcher.   (b) N-gram matcher.

(c) Evidence for subsumption.   (d) Evidence of equivalence.

Fig. 8. Effect on the posterior probabilities using particular evidence on different prior probabilities.

where the joint probability $P((f(m) = y \cap \delta(m) = x) \in B)$, $B \in [0,1]$ is computed with a double integral over the estimated density function (derived using KDE) $\hat{f}(y, x)$ as follows:

$$P(f(m) = y \cap \delta(m) = x) = \int \int_B \hat{f}(y, x) \, dx \, dy \qquad (8)$$

The resulting probability, using Eq. 8, can be seen as the area under the surface conditioned on the event $[a - \epsilon \le y \le a + \epsilon, c - \epsilon \le x \le c + \epsilon]$, where ϵ is a small positive number. $P(\delta(m) = x)$ in Eq. 7, is the marginal probability. We use the computed probability using Eq. 7 as the likelihood term required by Eq. 6.

For completeness, $P(f(m) = y)$ in Eq. 6 is a normalization factor to sum the probabilities to unity. This is the marginal probability denoted by $\int_{-\infty}^{\infty} f_Y(y \mid X = x) f_X(x) \, dx$.

Finally, $P(\delta(m) = x \mid f(m) = y)$ denotes the posterior probability that a value produced by a mapping m will be correct given an observed mapping fitness value y.

The Bayesian updating methodology described above can underpin the uniform and consistent assimilation of different types of evidence to yield judgements on the correctness/quality of the individual artefacts involved in a data integration life-cycle, i.e., matchings and mappings. Assimilation of new pieces of evidence leads to updates to the prior d.o.b.s in these artefacts, which can potentially be propagated to more complex artefacts or other phases in the life-cycle. Thus, the d.o.b.s in matching equivalences are propagated to the mapping generation process, which now uses those d.o.b.s as input rather than similarity scores as in most of the literature on this topic. Similarly, d.o.b.s on mapping correctness can be used as priors in an improvement phase that assimilate user feedback on mapping results. Enabling this principled propagation over many phases of a pay-as-you-go data integration process is a major contribution of this paper.

We observe that Bayesian updating, as such, is not computationally expensive but, of course, the construction of the likelihoods, which is essentially a training/induction step, could be, as it is involves labelling. In a real-world application where a specific concern leads to the generation of a specific training set, one would appeal to sampling theory in order to avoid an unnecessarily large training set. This, of course, need not be large. It rather needs to be representative of the underlying intended sources. So, this induction step, albeit relatively expensive, may need to be done once but possibly not again. Unless, of course, the population from which the sample was drawn changes significantly and irreversibly.

5   Experimental Evaluation

The evaluation of our approach is based on the idea of emulating the judgements produced by human experts in the presence of different kinds of evidence as the latter emerge from an automated data integration cycle. The collected judgements derived from experts were then compared with the judgements derived by the Bayesian updating approach (as discussed in Sects. 3 and 4).

This section describes our experimental evaluation, which had the following goals: (a) to compare how well the Bayesian assimilation of syntactic evidence alone performs against the aggregation of syntactic evidence followed by a predefined function (viz., average), which is commonly used in existing matching systems [3, 32]; (b) to ascertain whether the incorporation of semantic evidence can improve on judgements on the equivalence of constructs obtained through syntactic matching alone; (c) to ascertain whether the derived d.o.b.s on mapping correctness are consistent with the aggregated testimonies from human experts against the computed mapping degrees of correctness given a ground truth; and (d) to compare the d.o.b.s on mappings obtained using the Bayesian approach with the mapping correctness scores using the ground truth.

5.1   Experimental Setup

Use of Expert Testimonies. To evaluate the application of Bayes's theorem for assimilating different kinds of evidence, the experimental evaluation was grounded on the rational decisions made by human experts on data integration and ontology alignment when judging whether a pair of constructs is postulated to be equivalent given both syntactic and semantic evidence, and postulating whether a mapping expression will produce correct values, as construed in this paper. Fifteen human experts were asked (through surveys) to judge the correctness of matches and mapping expressions and their judgements were compared to the judgements obtained through the use of our methodology. By *experts*, we mean professionals in data integration.

Deriving Expert d.o.b.s in the Matching Stage. In the experiments investigating matching, a set of pairs of constructs from different LD ontologies was collected, making sure that different combinations of syntactic and semantic evidence (as in Table 3) were present or absent. To obtain testimonies from the human experts, a survey was designed based on the collected set of pairs of constructs, asking the experts to make judgements on the equivalence of such pairs. Testimonies were recorded on a discretization scale [6], as follows: {Definitely equivalent} mapped to a d.o.b. of 1.0; {Tending towards being equivalent} mapped to a d.o.b. of 0.75; {Do not know} mapped to a d.o.b. of 0.5; {Tending towards being not-equivalent} mapped to a d.o.b. of 0.25; and {Definitely not-equivalent} mapped to a d.o.b. of 0. By observing different pairs of constructs from real ontologies, approximately 40 common combinations of syntactic and semantic evidence have been identified. For each combination, a question was designed to obtain individual testimonies from each responder. Individual testimonies from each question were aggregated using a weighted average, based on the confidence assigned to each item [6]. The aggregated d.o.b.s obtained from the survey are treated as an approximation of the experts' confidence on equivalence of constructs given certain pieces of syntactic and semantic evidence and act as a gold standard.

Deriving Expert d.o.b.s in the Mapping Stage. Similarly, for the mapping experiments, a set of mappings and their results were collected and presented to human experts in order to obtain individual testimonies on the correctness of mapping results. The mappings were derived by integrating two real-world schemas from the music domain, viz., Jamendo[9] and Magnatune[10], using DSToolkit [13]. An on-line survey, consisting of a set of mapping expressions written in SPARQL and a sample of the corresponding result tuples, was delivered to expert users who were asked to postulate how likely it was that a value in the tuples produced by that mapping would be correct. We used the following discretization scale: {Definitely correct} mapped to a d.o.b. of 1.0; {Tending towards being correct} mapped to a d.o.b. of 0.75; {Do not know/Partially correct} mapped to a d.o.b. of 0.5; {Tending towards being incorrect} mapped to a d.o.b. of 0.25; and {Definitely incorrect} mapped to a d.o.b. of 0. For each question, we aggregated the individual testimonies using an average. We treat the aggregated testimonies as an approximation of a human-derived d.o.b. on mapping correctness.

---

[9] http://dbtune.org/jamendo/
[10] http://dbtune.org/magnatune/

Datasets for the Matching Stage. For the purposes of the matching experiment, the Bayesian technique was evaluated over the class hierarchies of ontologies made available by the OAEI (Conference Track). These have been designed independently but they all belong to the domain of conference organisation. Note also that these ontologies share no semantic relations between them. Since our technique assumes such relations for use as semantic evidence, we made explicit some of these cross-ontology semantic relations using BLOOMS[11], a system for discovering rdfs:subClassOf and owl:equivalentClass relations between LD ontologies [16]. We note that the contributions reported in this paper are independent of BLOOMS, in that they can be used regardless of the sources of semantic annotations. We found that, as it currently stands, the LOD cloud still lacks the abundance of cross-ontology links at the conceptual level that is implied by the vision of a Semantic Web. The results reported in this paper consider a single pair of ontologies from the conference track, viz., `ekaw` (denoted by S) and `conference` (denoted by T).

Datasets for the Mapping Stage. The set of mappings used in the experiments was derived using schemas from the music domain. In particular, we used three schemas: Magnatune (as a source schema), Jamendo (as a source schema), and DBTune (as the target schema). Magnatune is an online music streaming service which offers an online music catalog. Jamendo is a linked open data repository. DBTune is an ontology that describes music artists, records, tracks, and performances. The schemas are depicted in Fig. 9.

Expectation Matrix. Given a pair of classes from the class hierarchies of the input ontologies and given the available kinds of evidence, both syntactic and semantic, a d.o.b. was assigned for each pair on the basis of the experts' testimonies. More formally, we constructed a $n \times m$ structure referred to from now on as the *expectation matrix* and denoted by M_{exp}, where $n = |S|$ and $m = |T|$. The element e_{jk} in the jth row and the kth column of M_{exp} denotes the d.o.b. derived from the expert survey between the jth construct in S and the kth construct in T according to the pieces of evidence present or absent. Similarly, we constructed a vector $e = e_1, \ldots, e_n$, $n = |M|$, where the element e_i denotes the d.o.b. derived from the expert survey for the mapping $m_i \in M$, and the vector $b = b_1, \ldots, b_n$, $n = |M|$, where the element b_i denotes the d.o.b. derived by the Bayesian approach for the mapping $m_i \in M$.

Evaluation Metric. Let $p_1, p_2, \ldots, p_n$ be the d.o.b.s derived for each pair of classes from the ontologies by either the average aggregation scheme or the Bayesian assimilation, and let $a_1, a_2, \ldots, a_n$ be the corresponding d.o.b.s in the expectation matrix just described. In the same way, let $p_1, p_2, \ldots, p_n$ be the d.o.b.s for each mapping $m_i \in M$ by the Bayesian approach, and let $a_1, a_2, \ldots, a_n$ be the corresponding d.o.b.s from each mapping $m_i \in M$ by the experts' testimonies. We compute the mean-absolute error, $MAE = (|p_1 - a_1| + \ldots + |p_i -$

---

[11] BLOOMS was configured with a high threshold, viz., >0.8.

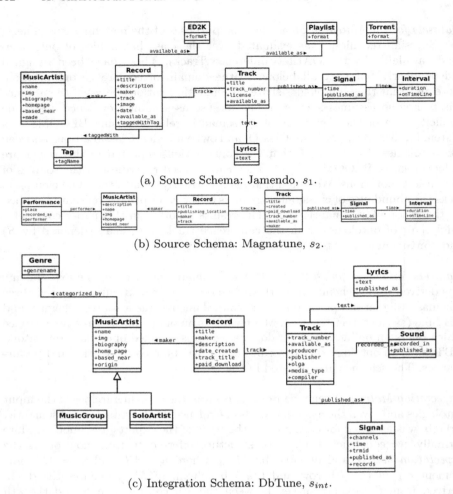

(a) Source Schema: Jamendo, s_1.

(b) Source Schema: Magnatune, s_2.

(c) Integration Schema: DbTune, s_{int}.

Fig. 9. Schemas for deriving mappings.

$a_n|)/n$ where $|p_i - a_i|$ is the *individual error* of the i-th pair and n is the total number of such errors. We also compute the correlation coefficient ρ between mapping d.o.b.s and mapping correctness scores, and between mapping d.o.b.s and aggregated experts' testimonies, $\rho_{X,Y} = cov(X, Y)/\delta X \delta Y$, where X is the set of mapping d.o.b.s and Y is either the set of mapping correctness scores or the set of aggregated experts' testimonies.

5.2 Experimental Design

Traditional matching approaches (e.g., COMA [1]) exploit different pieces of evidence, mostly from string-based matchers, to assess the similarity between constructs in ontologies or in database schemas. Such approaches combine similarity scores computed independently, typically using averages. For the matching evaluation, the antagonist to our Bayesian approach is a process that independently runs matchers ng and ed on the local-names of classes from ontologies S and T, and produces an average of the similarity scores. The aggregated result of this computation is a matrix M_{avg}. The next step is to measure how close the derived predictions are to the d.o.b.s obtained by the experts' testimonies. In doing so, we used MAE as the performance measure since it does not exaggerate the effect of outliers [15]. The result from computing the error between M_{avg} and the expectation matrix M_{exp} is denoted by δ_{avg}.

Similarly, the Bayesian assimilation technique (as described in Sect. 4) was used (instead of an average) to assimilate the evidence computed by the string-based matchers on pairs of local-names. The result of this computation is a matrix M_{syn}, where $n = |S|$ and $m = |T|$. The element e_{jk} in the jth row and the kth column of M_{syn} denotes the posterior probability $P^{(syn)}$ between the jth class in S and the kth class in T according to the syntactic evidence derived from the string-based matchers ed and ng. The next step is to measure how close the predictions from M_{syn} are to the expectation matrix M_{exp}. The result is denoted by δ_{syn}.

To assess whether semantic evidence can improve on judgements on the equivalence of constructs that use averaging alone to aggregate syntactic evidence, we first used BLOOMS [16] to make explicit the cross-ontology semantic relations and used this as semantic evidence. In the light of this new evidence, the Bayesian assimilation technique updates the posterior probabilities $P^{(syn)}$ for each pair of classes in M_{syn} accordingly. The result of this process is a new matrix $M_{syn,sem}$ with the same dimensions as M_{syn}, where, the posterior probabilities for the elements e_{jk} reflect both syntactic and semantic evidence, $P^{(syn,sem)}$. Again we denote by $\delta_{syn,sem}$ the error calculated between $M_{syn,sem}$ and the expectation matrix M_{exp}. Finally, to complete the evaluation, the individual absolute errors used for the calculation of δ_{avg}, δ_{syn}, and $\delta_{syn,sem}$ have been examined.

To evaluate the derived d.o.b.s on mapping correctness, we compared the resulting d.o.b.s against two measures: *aggregated d.o.b.s* that were obtained from testimonies from human experts, and *overall mapping correctness* with respect to an available ground truth. In both cases, we observed whether a derived d.o.b. for a mapping by the Bayesian approach is consistent with the d.o.b. estimated from human experts, and an estimated mapping correctness score given a ground truth. We would expect that a low d.o.b., e.g., lower than 0.1 by the Bayesian approach, should relate to a low d.o.b. obtained from either human experts or from an observed mapping correctness score. In contrast, a high d.o.b., e.g., greater than 0.6, should likewise relate to a high d.o.b. derived from human testimonies and from an estimated mapping correctness score. We use MAE to estimate the overall error between the d.o.b.s by the Bayesian and the experts' testimonies, and the computed similarity given an available ground truth obtained from a Benchmark.

5.3 Results and Discussion

In experiments 1–3, individual errors are correlated against the expected value (from experts' testimonies).

Exp. 1 – Matching: AVG scheme vs. Bayesian Syntactic. The MAE error computed for the average aggregation scheme against the expectation matrix was $\delta_{avg} = 0.1079$ whereas the error as a result of assimilating syntactic evidence using the Bayesian technique was $\delta_{syn} = 0.0698$. To further understand the difference in errors, we measured the individual *absolute* errors that fall into each of four regions of interest as these are shown in Fig. 10(a). They correspond to the following minimum bounding rectangles, resp., Region 1 lies below the $y = x$ error line where AVG error $>>$ Bayesian error and is the rectangle defined by $y = 0.2$; Region 2 lies above the $y = x$ error line where AVG error $<<$ Bayesian error and is the rectangle defined by $x = 0.2$; Region 3 lies below the $y = x$ error line where AVG error $>$ Bayesian error and is the rectangle defined by $y > 0.2$; and Region 4 lies above the $y = x$ error line where AVG error $<$ Bayesian error and is the rectangle defined by $x > 0.2$. We note that the larger the cardinality of Region 1, the more significant is the impact of using semantic annotations as we propose.

For the traditional aggregation scheme that produced M_{avg} we counted 3833 matches with individual errors greater than the analogous individual errors derived by the Bayesian technique that produced M_{syn}. The use of Bayesian aggregation significantly outperformed (i.e., has smaller individual errors than) the use of AVG aggregation scheme for 87.49% of the total. Table 5 summarises the results for each region showing how many individual errors are located in each of the regions of interest in both absolute terms and relative to the total.

Exp. 2 – Matching: AVG scheme vs. Bayesian Syn. & Sem. To evaluate our hypothesis that semantic annotations can improve outcomes we compared the aggregated errors denoted by δ_{avg} and $\delta_{syn,sem}$. The mean absolute error $\delta_{syn,sem} = 0.1259$ is lower than $\delta_{avg} = 0.1942$ with a difference of 0.0683. Figure 10(b) plots the individual errors for pairs of classes that have some semantic relation between them. We are interested on cases where the individual errors for the Bayesian technique are smaller than the AVG scheme. In particular, the points that lie mostly between 0.1 and 0.3 on the x-axis and below the $y = x$ error line. For 71.43% of the total matches that have some semantic evidence the Bayesian technique produces results closer to the testimonies, with individual errors that mostly lie in that region. Table 6 summarises the results for each region showing how many individual errors are located in each of the regions of interest in both absolute terms and relative to the total.

Table 5. AVG scheme vs. Bayesian syntactic.

No.	Region	Count	Perc. (%)
1	$R_{avg \gg B_{syn}}$	3833	87.49
2	$R_{avg \ll B_{syn}}$	215	4.90
3	$R_{avg > B_{syn}}$	31	0.70
4	$R_{avg < B_{syn}}$	302	6.89

Table 6. AVG scheme vs. Bayesian syntactic & semantic.

No.	Region	Count	Perc. (%)
1	$R_{avg \gg B_{syn,sem}}$	125	71.43
2	$R_{avg \ll B_{syn,sem}}$	43	24.57
3	$R_{avg > B_{syn,sem}}$	2	1.14
4	$R_{avg < B_{syn,sem}}$	5	2.85

Table 7. Bayesian syntactic vs. Bayesian syntactic & semantic.

No.	Region	Count	Perc. (%)
1	$R_{B_{syn} \gg B_{syn,sem}}$	124	89.21
2	$R_{B_{syn} \ll B_{syn,sem}}$	9	6.48
3	$R_{B_{syn} > B_{syn,sem}}$	5	3.60
4	$R_{B_{syn} < B_{syn,sem}}$	1	0.72

Exp. 3 – Matching: Bayesian Syn. vs. Bayesian Syn. & Sem. Similarly to Exp.2, we compared the aggregated errors denoted by δ_{syn} and $\delta_{syn,sem}$ considering only individual errors that have some semantic evidence. Again in this case $\delta_{syn,sem} = 0.1259$ is closer to the expectation matrix than $\delta_{syn} = 0.2768$ with a difference of 0.1509. The results of this experiment are summarised in Table 7. The points of interest in this experiment are the ones where the individual errors for $B_{syn,sem}$, that considers both syntactic and semantic evidence, are smaller than B_{syn}. For 89.21% of the total matches discovered, that have some semantic evidence, $B_{syn,sem}$ outperforms the configuration of the Bayesian scheme that utilises syntactic evidence alone, i.e., B_{syn}.

For the mapping generation case, we focus on the correlation between the aggregated d.o.b.s from experts' testimonies against the overall mapping correctness score, derived using ground truth, as well as with the d.o.b.s derived by assimilating mapping generation evidence (i.e., fitness values) using the Bayesian approach.

Exp. 4 – Mapping Generation: Bayesian d.o.b.s vs. Observed Mapping Correctness. For each mapping in the integration, we observed the d.o.b. derived by the Bayesian approach (x), and the mapping correctness score (y) using an available ground truth. We correlate these two measures in a scatter plot depicted in Fig. 11(a). Here, we can observe that the Bayesian approach (x-axis) is being more optimistic than the computed similarity using an available ground truth. One possible reason for this is that the ground truth is inherently rigorous in the sense that it does not allow for misleading interpretations of the actual data. Thus, mapping correctness tends to be lower than the derived d.o.b.s. Comparing the two measures, it can be seen that there is a positive correlation between the Bayesian d.o.b.s and the mapping similarities as, for most cases, a low d.o.b.

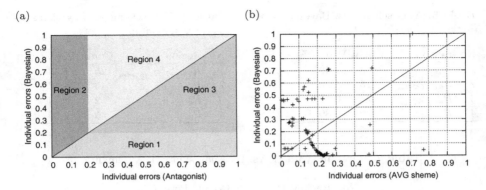

Fig. 10. (a) Shows the regions of interest, (b) Individual errors Bayesian against AVG scheme.

correlates with a low mapping correctness. Similarly, a high d.o.b. relates to a high similarity score. The computed MAE for the Bayesian d.o.b.s against the mapping correctness was $\delta = 0.1274$. The correlation coefficient between the Bayesian d.o.b.s and the mapping correctness score is 0.92. Furthermore, the computed MAE for the Bayesian d.o.b.s against the mapping correctness was $\delta = 0.1296$.

Exp. 5 – Mapping Generation: Bayesian d.o.b.s vs. Experts Testimonies. As in Experiment 4, here we correlate the d.o.b. derived by the Bayesian approach with the aggregated testimonies from experts. Figure 11(b) depicts this correlation. Here, we observe that the experts' testimony is slightly more optimistic than the d.o.b.s derived by the Bayesian approach. Moreover, we observe that there is a positive correlation, i.e., low d.o.b.s are correlated to low d.o.b.s by experts' testimonies, whereas high d.o.b.s are correlated to high d.o.b.s from experts' testimonies. The correlation coefficient between the Bayesian d.o.b.s and the experts testimonies is strong but slightly lower than in Exp. 4, possibly due to inevitable subjectivity, albeit reduced by expertise, in human judgements. The computed MAE for the Bayesian d.o.b.s against the testimony from experts was $\delta = 0.1113$.

We also show the individual errors between the aggregated testimonies from experts against the overall mapping correctness score, using ground truth. This is depicted in Fig. 11(c). Here, we observe that in most cases, the individual errors are low, i.e., <0.2. This may suggest that both techniques derive closely related measures for individual mappings.

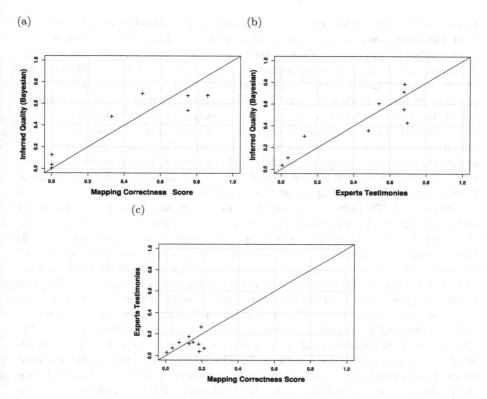

Fig. 11. (a) Inferred quality vs. observed quality, (b) Inferred quality vs. experts' testimonies, (c) Individual errors overall mapping correctness score vs aggregated experts testimonies.

6   Related Work

Automatic techniques for bootstrapping a data integration system offers opportunities for on-demand approximate integrations [31]. This approximation arises from the different kinds of uncertainty propagated throughout the process of integration. In the context of LD sources, automatic *schema extraction* techniques [5] are used to approximate the structure of the sources which is not strictly enforced. *Matching* techniques are likely to be uncertain due to the robustness of the matching techniques where the associations discovered between the sources require selection and grouping to inform the generation of mappings. This uncertainty on the match results is propagated throughout to *mapping generation* [2] influencing the ability to produce correct results. In this paper, we make the case that the effects introduced by the inherited uncertainty can be better understood by assimilating different forms of evidence in a principled, uniform manner throughout the integration processes. We then position this work in relation to other proposals that are concerned with these challenges.

Reacting to Different Pieces of Evidence for Matching. A variety of strategies have been proposed in the literature for solving the problem of combining different pieces of evidence about matches, some examples are: average, weighted average, min, max and sigmoid functions [25]. However, it falls on users to tune or select the appropriate aggregation method manually according to the problem in hand. In contrast, the Bayesian assimilation of evidence technique can be used as an alternative aggregation strategy for assimilating any piece of evidence, complementing typical aggregation strategies used by state-of-the-art schema and ontology matching systems [3,27,32]. When the appropriate probability distributions are made available, the approach presented in this paper can be used as a generic aggregation strategy that presents results in terms of d.o.b.s, rather than building on matcher-specific metrics.

Sabou et al. [28] presented an ontology matching paradigm that makes use of additional external background knowledge that is made available from ontologies from the Semantic Web. The proposal in our paper makes use of additional semantic annotations from LD ontologies as evidence with the aim of improving the decision making of different matchers that mostly work on syntax. Approaches for discovering semantic relations from ontologies e.g., [29] can be used to provide input to our Bayesian approaches to further improve the accuracy, thus improving the decision making of matching approaches. The uncertainty in the decisions made by different *matchers* has also been observed in [22], where a similarity matrix that describes the outcome of some matcher is modelled as two probability distributions. An alternative statistical analysis is used to model the similarity scores distribution returned by each matcher that uses the parametric beta-distribution to estimate the underlying probability. The proposal in our paper, however, makes no assumptions about the shape or parameters of the underlying distribution, and uses a non-parametric statistical analysis technique, based on kernel density estimation, to approximate the probability distributions for each matcher using the sampled data.

We observe that the antagonist in our matching experiments (i.e., taking the average of a collection of independently-produced similarity scores) is the de facto standard for schema matching.

Uncertainty Management in Mapping Generation. Dong *et al.* proposed an approach to manage uncertainty in data integration by introducing the concept of probabilistic schema mappings, in which a probability is attached to each generated mapping. This probability is derived from a probability mass function on the fraction of attributes from a source schema that conform to attributes in a mediated schema [7]. The assigned probability is used to produce a result consisting of the top-k tuples during the query evaluation process. In our work the probability assigned to each mapping denotes the d.o.b. that a tuple produced by a mapping is likely to be correct, whereas in [7] the assigned probability denotes the d.o.b that a mapping is correct among the mappings that describe the same source and target concept. In addition, we are not restricted to one-to-one mappings, as we also deal with one-to-many relationships. [21] assumes the existence of a set of matches annotated with probabilities to present a data integration process that annotates mappings with probabilities. In relation to mappings, they use a discrete value to denote the semantic relationship between constructs, whereas in our work we assign a degree of correctness in the continuous interval $[0, 1]$. We do not simply assume the existence of probabilities, instead we have described a systematic methodology for deriving them. In another study, van Keulen [35], proposes a probabilistic approach to deal with uncertainty in data cleaning, mapping and information extraction approaches. Here, uncertainty is model as random events representing assertions on data instances, i.e., whether two data instances relates to the same real-world object or not. In contrast, our approach deals with uncertainty in postulating syntactic and semantic equivalence between schema constructs from different sources.

We observe that there is no comparable antagonist for our mapping generation experiments insofar as work on mapping generation has mostly stemmed from the data exchange literature and hence has focussed on generating mappings that can be used to materialize core solutions to the data exchange problem, whereas our contribution aims at producing a quantification of the uncertainty associated with automatic mapping generation that closely correlates with the corresponding expert judgements.

Most experimental work on automating data integration techniques is by and large incomparable with ours because, so far, their primary intent has been on evaluating a point solution (i.e., a technique that applies to a single stage, such as matching, or mapping generation, of the end-to-end approach) whereas one of our main goals has been to evaluate a cross-stage technique, i.e., one of our contributions is to show how the quantified uncertainty resulting from the matching stage influences the quantified uncertainty associated with the generated mappings in the subsequent stage.

To the best of our knowledge, our work is the first attempt to evaluate the techniques on their ability to correlate closely to the corresponding judgement of experts. This is as ambitious as it is onerous and strongly suggests that future work is needed to collect more data points and ascertain the true robustness of our experimental results.

7   Conclusions

The WoD can be seen as vibrant but challenging: vibrant because there are numerous publishers making valuable data sets available for public use; challenging because of inconsistent practises and terminologies in a setting that is something of a free-for-all. In this context, it is perhaps easier to be a publisher than a consumer. As a result, there is a need for tools and techniques to support effective analysis, linking and integration in the web of data [26].[12]

---

[12] We observe once more that, in this paper, the experiments have only used LD datasets but dataspaces are meant to be model-agnostic and, in particular, DSToolkit is. DSToolkit is no longer being actively developed but requests for access to the sources can be sent to the second author. The datasets used are publicly available in the LOD cloud.

The challenging environment means: (i) that there are many different sources of evidence on which to build; (ii) that there is a need to make the most of the available evidence; and (iii) that it is not necessarily easy to do (ii). This paper has described a well-founded approach to combining multiple sources of evidence of relevance to matching and mapping, namely similarity scores from several syntactic matchers, semantic annotations, and mapping generation evidence in the form of fitness values. The main finding from our experimental results is confirmation that the contributed Bayesian approach can be used as a generic approach of assimilating different kinds of evidence that are likely to emerge throughout an automated integration process, in ways that reflect the opinions of human integration experts.

Acknowledgments. Fernando R. Sanchez S. is supported by a grant from the Mexican National Council for Science and Technology (CONACyT).

References

1. Aumueller, D., Do, H.H., Massmann, S., Rahm, E.: Schema and ontology matching with COMA++. In: SIGMOD Conference, pp. 906–908 (2005)
2. Belhajjame, K., Paton, N.W., Embury, S.M., Fernandes, A.A.A., Hedeler, C.: Incrementally improving dataspaces based on user feedback. Inf. Syst. **38**(5), 656–687 (2013)
3. Bernstein, P., Madhavan, J., Rahm, E.: Generic schema matching, ten years later. Proc. VLDB Endow. **4**(11), 695–701 (2011)
4. Bowman, A.W., Azzalini, A.: Applied Smoothing Techniques for Data Analysis: The Kernel Approach with S-Plus Illustrations. OUP, Oxford (1997)
5. Christodoulou, K., Paton, N.W., Fernandes, A.A.A.: Structure inference for linked data sources using clustering. In: Hameurlain, A., Küng, J., Wagner, R., Bianchini, D., De Antonellis, V., De Virgilio, R. (eds.) Transactions on Large-Scale Data- and Knowledge-Centered Systems XIX. LNCS, vol. 8990, pp. 1–25. Springer, Heidelberg (2015). https://doi.org/10.1007/978-3-662-46562-2_1
6. de Vaus, D.: Surveys in Social Research: Research Methods/Sociology. Taylor & Francis, London (2002)
7. Dong, X.L., Halevy, A.Y., Yu, C.: Data integration with uncertainty. VLDB J. **18**(2), 469–500 (2009)
8. Guo, C., Hedeler, C., Paton, N.W., Fernandes, A.A.A.: EvoMatch: an evolutionary algorithm for inferring schematic correspondences. In: Hameurlain, A., Küng, J., Wagner, R. (eds.) Transactions on Large-Scale Data- and Knowledge-Centered Systems XII. LNCS, vol. 8320, pp. 1–26. Springer, Heidelberg (2013). https://doi.org/10.1007/978-3-642-45315-1_1
9. Guo, C., Hedeler, C., Paton, N.W., Fernandes, A.A.A.: MatchBench: benchmarking schema matching algorithms for schematic correspondences. In: Gottlob, G., Grasso, G., Olteanu, D., Schallhart, C. (eds.) BNCOD 2013. LNCS, vol. 7968, pp. 92–106. Springer, Heidelberg (2013). https://doi.org/10.1007/978-3-642-39467-6_11
10. Halevy, A.Y.: Why your data won't mix: semantic heterogeneity. ACM Queue **3**(8), 50–58 (2005)

11. Halevy, A.Y., Franklin, M.J., Maier, D.: Principles of dataspace systems. In: PODS, pp. 1–9 (2006)
12. Halevy, A.Y., Rajaraman, A., Ordille, J.J.: Data integration: the teenage years. In: VLDB, pp. 9–16 (2006)
13. Hedeler, C., et al.: DSToolkit: an architecture for flexible dataspace management. In: Hameurlain, A., Küng, J., Wagner, R. (eds.) Transactions on Large-Scale Data- and Knowledge-Centered Systems V. LNCS, vol. 7100, pp. 126–157. Springer, Heidelberg (2012). https://doi.org/10.1007/978-3-642-28148-8_6
14. Hedeler, C., Belhajjame, K., Paton, N.W., Campi, A., Fernandes, A.A.A., Embury, S.M.: Chapter 7: dataspaces. In: Ceri, S., Brambilla, M. (eds.) Search Computing. LNCS, vol. 5950, pp. 114–134. Springer, Heidelberg (2010). https://doi.org/10.1007/978-3-642-12310-8_7
15. Hyndman, R.J., Koehler, A.B.: Another look at measures of forecast accuracy. IJF 22(4), 679–688 (2006)
16. Jain, P., Hitzler, P., Sheth, A.P., Verma, K., Yeh, P.Z.: Ontology alignment for linked open data. In: Patel-Schneider, P.F., et al. (eds.) ISWC 2010. LNCS, vol. 6496, pp. 402–417. Springer, Heidelberg (2010). https://doi.org/10.1007/978-3-642-17746-0_26
17. Kim, W., Seo, J.: Classifying schematic and data heterogeneity in multidatabase systems. IEEE Comput. 24(12), 12–18 (1991)
18. Kuicheu, N.C., Wang, N., Fanzou Tchuissang, G.N., Xu, D., Dai, G., Siewe, F.: Managing uncertain mediated schema and semantic mappings automatically in dataspace support platforms. Comput. Inform. 32(1), 175–202 (2013)
19. Lenzerini, M.: Data integration: a theoretical perspective. In: PODS, pp. 233–246 (2002)
20. Madhavan, J., et al.: Web-scale data integration: you can only afford to pay as you go. In: CIDR, pp. 342–350 (2007)
21. Magnani, M., Montesi, D.: Uncertainty in data integration: current approaches and open problems. In: Proceedings of the First International VLDB Workshop on Management of Uncertain Data in Conjunction with VLDB 2007, Vienna, Austria, 24 September 2007, pp. 18–32 (2007)
22. Marie, A., Gal, A.: Managing uncertainty in schema matcher ensembles. In: Prade, H., Subrahmanian, V.S. (eds.) SUM 2007. LNCS (LNAI), vol. 4772, pp. 60–73. Springer, Heidelberg (2007). https://doi.org/10.1007/978-3-540-75410-7_5
23. Papoulis, A.: Probability, Random Variables and Stochastic Processes, 3rd edn. McGraw-Hill Companies, New York (1991)
24. Paton, N.W., Belhajjame, K., Embury, S.M., Fernandes, A.A.A., Maskat, R.: Pay-as-you-go data integration: experiences and recurring themes. In: Freivalds, R.M., Engels, G., Catania, B. (eds.) SOFSEM 2016. LNCS, vol. 9587, pp. 81–92. Springer, Heidelberg (2016). https://doi.org/10.1007/978-3-662-49192-8_7
25. Peukert, E., Maßmann, S., König, K.: Comparing similarity combination methods for schema matching. In: GI Jahrestagung, no. 1, pp. 692–701 (2010)
26. Polleres, A., Hogan, A., Harth, A., Decker, S.: Can we ever catch up with the web? Semant. Web 1(1–2), 45–52 (2010)
27. Rahm, E., Bernstein, P.A.: A survey of approaches to automatic schema matching. VLDB J. 10(4), 334–350 (2001)
28. Sabou, M., d'Aquin, M., Motta, E.: Exploring the semantic web as background knowledge for ontology matching. J. Data Semant. 11, 156–190 (2008)

29. Sabou, M., d'Aquin, M., Motta, E.: SCARLET: Semantic relation discovery by harvesting online ontologies. In: Bechhofer, S., Hauswirth, M., Hoffmann, J., Koubarakis, M. (eds.) ESWC 2008. LNCS, vol. 5021, pp. 854–858. Springer, Heidelberg (2008). https://doi.org/10.1007/978-3-540-68234-9_72
30. Das Sarma, A., Dong, X., Halevy, A.Y.: Bootstrapping pay-as-you-go data integration systems. In: SIGMOD Conference, pp. 861–874 (2008)
31. Sarma, A.D., Dong, X.L., Halevy, A.Y.: Uncertainty in data integration and dataspace support platforms. In: Bellahsene, Z., Bonifati, A., Rahm, E. (eds.) Schema Matching and Mapping. DCSA, pp. 75–108. Springer, Heidelberg (2011). https://doi.org/10.1007/978-3-642-16518-4_4
32. Shvaiko, P., Euzenat, J.: Ontology matching: state of the art and future challenges. IEEE Trans. Knowl. Data Eng. **25**(1), 158–176 (2013)
33. Silverman, B.W.: Density Estimation for Statistics and Data Analysis. Chapman & Hall, London (1986)
34. Spragins, J.: A note on the iterative application of Bayes' rule. IEEE Trans. Inf. Theory **11**(4), 544–549 (2006)
35. van Keulen, M.: Managing uncertainty: the road towards better data interoperability. IT - Inf. Technol. **54**(3), 138–146 (2012)

A Comprehensive Approach for Designing Business-Intelligence Solutions with Multi-agent Systems in Distributed Environments

Karima Qayumi[1]([envelope]) and Alex Norta[2]([envelope])

[1] School of Digital Technologies, Tallinn University,
Narva Mnt 29, 10120 Tallinn, Estonia
karima.qayumi@gmail.com
[2] Large-Scale-Systems Group, Tallinn University of Technology,
Akadeemia Tee 15a, 12616 Tallinn, Estonia
alex.norta.phd@ieee.org

Abstract. Multi-agent systems (MAS) are an active research area of system engineering to deal with the complexity of distributed systems. Due to the complexity of business-intelligence (BI) generation in a distributed environment, the adaptation of such system is diverse due to integrated MAS and distributed data mining (DDM) technologies. Bringing these two frameworks together in the content of BI-systems poses challenges during the analysis, design, and test in the development life-cycle. The development processes of such complex systems demand a comprehensive methodology to systematically guide and support developers through the various stages of BI-system life-cycles. In the context of agent-based system engineering, several agent-oriented methodologies exist. Deploying the most suitable methodology is another challenge for developers. In this paper, we develop an exemplar of MAS-based BI-system called BI-MAS with comprehensive designing steps as a running case. For demonstrating the new approach, first we consider an evaluation process to find suitable agent-oriented methodologies. Second, we apply the selected methodologies in analyzing and designing concepts for BI-MAS life-cycles. Finally, we demonstrate a new approach of verification and validation processes for BI-MAS life-cycles.

Keywords: Business-intelligence (BI) · Distributed data mining (DDM)
Multi-agent system (MAS) · Agent-oriented modeling (AOM)

1 Introduction

Business-intelligence (BI) is a modern management support that includes users, distributed data mining (DDM) processes, intelligence tools, information management, and analysis processes in order to improve decision-making and business performance [1–3]. Agent technologies, or multi-agent systems (MAS) [1, 4, 5] are promoted as an emerging technology that facilitates the design, implementation, and maintenance of distributed systems. DDM [6, 7] originates from the need for mining intelligence over decentralized data sources. Furthermore, DDM is known as one of the latest solutions,

© Springer-Verlag GmbH Germany, part of Springer Nature 2018
A. Hameurlain and R. Wagner (Eds.): TLDKS XXXVII, LNCS 10940, pp. 113–150, 2018.
https://doi.org/10.1007/978-3-662-57932-9_4

or procedures to reduce massive data-discovery problems in highly distributed environments [1].

Researchers apply MAS together with DDM for reducing the complexities of BI-systems in distributed environments [1, 9, 10]. Additionally, literature also comprises several types of agent-based architectures and frameworks either in the content of DDM, or BI-systems in [11–16]. Our studies discover that there exists a lack in deployment of agent-oriented methodology in these mentioned architectures and frameworks. Conversely, literature presents numerous types of agent-oriented methodologies in [17–20, 29–35].

In the context of software engineering [2], each model of the designing phase must describe a specific aspect of the system under consideration. In fact, the designing processes of agent-based BI-systems require either an applicable agent-oriented methodology, or a significant approach to capture requirement specifications and translate them for the development process. The development process of BI-MAS that comprise different types of agents in a distributed environment with the ability to communicate, discover and access data from multiple sites, requires a comprehensive methodology [3]. The main challenges for developing processes of BI-MAS are the assigning of agents to perform tasks in parallel and the management of collaborations and cooperation processes in complex applications [11, 21]. In such complex systems, developers need a unified agent-oriented methodology for the entire life-cycle of agent-based BI-systems.

In this paper, we address the current gap in the state-of-the-art for developing a process of BI-MAS by answering the research question of how to develop a designing approach for MAS-based BI-systems in distributed environments? To establish a separation of concerns, we elicit the following sub-questions.

RQ1. What evaluation is required to find applicable methods out of existing agent-oriented methodologies?
RQ2. What level of support do existing agent-oriented methodologies yield for developers to define a systematic way for the conceptualization of BI-MAS models?
RQ3. What types of methods and tools are demanded to consider the verification and validation (V&V) processes for proposed BI-MAS models?

The rest of the paper is structured as follows: Sect. 2 presents the related concepts of BI-MAS, challenges of traditional BI, and new BI-MAS features in business environments. Section 3 discusses existing agent-oriented methodologies and their evaluation results. Section 4 outlines the analysis and detailed design processes of the BI-MAS architecture by deploying combined agent-oriented methodologies. Section 5 comprises the mapping processes for BI-MAS models to a formalization using Colored Petri Nets (CPN) and the results from model checking that are explored from validating and the verification processes of the BI-MAS life-cycle. Section 6 describes the evaluation results together with a critical comparative discussion that compares the results of this paper against results from other research work. Finally, in Sect. 7 we conclude our paper with a summary of our research findings and suggestions for the future development of our research.

2 Related Concepts and Challenges

In last decades, research shows that developing BI-systems changes considerably due to different views of business owners, presentation of business-concepts, and the idea behind access to business data. To resolve the current challenges of BI-systems, research studies the combination of MAS and DDM technologies as a new-generation for such systems. In this section, we briefly discuss main components, current challenges of BI, and concept of agent-based BI-systems.

2.1 Business-Intelligence Systems

BI is a term that refers to technologies, applications, and practices for the collection, integration, analysis, and presentation of business information [22, 23]. BI-systems are a set of applications, technologies and tools for the transformation of raw data into meaningful and useful information in order to improve decisions and increase business performance [1, 24]. Therefore, BI refers to broad categories of applications and technologies that are used for corporate management, optimization of costumer relations, monitoring of business activities, data mining, reporting, planning, and decision-making support on all levels of managements [4]. The value of a BI-system for business is to provide adequate and reliable up-to-date information on various aspects of enterprise activities in an organization.

BI-systems comprise an integrated set of technologies and tools that contain several modules such as Extract, Transform and Load (ETL), data warehouse, Online Analytical Processing (OLAP), and tools for data mining and reporting [24, 25]. ETL is the set of processes relevant for the transformation, organization, and integration of loading data from numerous applications and systems into target systems, e.g., data warehouses. According to [4], a data warehouse is a subject oriented, integrated, time-variant, and non-volatile collection of data that provides generalized and consolidated data in multidimensional views. OLAP techniques use data warehouses designed for sophisticated enterprise BI-systems for the interactive and effective analysis of data in multidimensional spaces. Data-mining methods such as association, clustering, classification, prediction can be integrated with OLAP operations to enhance the interactive mining of knowledge from various data sources [4].

2.2 Current Challenges of BI-Systems

Recent discussions about BI issues include OLAP techniques, data mining, and data warehouses [5]. Business users rarely have real time access to data and work on historical data that are not updated regularly. Most BI-applications need an expert/specialist to run statistical reports, or data-mining processes to generate reports for business users [6]. According to [7], the challenges of analyzing and generating information are reflected with 27.4% when business users try to collect a single version of real-time fact from multiple data sources and systems. Furthermore, the management of information challenges is reported with 35.8% for delivering, self-service reporting and analyzing of data.

On the other hand, authors of [8] discover that existing BI-systems are deficient in three points. Firstly, the current BI-systems can only provide solutions for specified situation. Secondly, current BI-systems cannot deal with data from dynamic environments. Thirdly, the update speed is slow for BI-systems where the source code must be rewritten when a new requirement is added. When the size of data highly increases for existing BI-systems, this is another challenge for BI-analysts to react immediately to events as they occur. Therefore, real-time BI-systems (RTBI) [6] emerge to provide real-time tactical support for immediate enterprise actions in reaction to events that employ classic data warehousing for deriving information. Additionally, RTBI also need a comparison between present business events and historical patterns in order to automatically detect problems in distributed environments.

2.3   Concept of Agent-Based BI-Systems

An intelligent agent (or simply an agent) is a piece of software, or a computer system that performs services and gathers information autonomously [1, 5]. Agents need to display intelligence properties in order to perceive their environment and be autonomous for performing tasks on behalf of the users in heterogeneous environments [9]. Intelligence also improves the capability of an agent, while interacting with the context to perceive changes during knowledge exploration [10]. Hence, MAS provide an effective approach for coordination and cooperation among multiple units in complex distributed systems [11] and therefore, researchers incorporate MAS technology with data-mining algorithms for developing agent/based BI-systems [12]. Recent research of literature shows a trend for developing agent-based BI-systems in different domains such as e-commerce, supply chain management, resource allocation, intelligent production and so on [1, 28]. MAS are identified as a multiple role player in BI-systems, e.g., user behavior learning, customizing interaction information, and user notification when important events occur. Consequently, recent examples of agent-based BI-systems are reported about in [1, 4, 10, 27, 28, 54–56].

3   Designing Method

In this research, we propose a new designing solution for novel BI-systems that combine MAS and DDM technologies. The term of artifact is used to describe the high-level overview of BI-MAS components and therefore, the design-science research (DSR) [26] framework is applied for understanding, executing, and evaluating our proposed artifact. According to [13], for designing a new artifact, rigor is achieved by appropriately applying existing foundations and methodologies. With respect to DSR, first, we refer into nine well-known existing agent-oriented methodologies introduced in [17–20, 29–35]. Our studies discover that each of these methodologies has its own strengths and weaknesses, and respected coverage phases that are limited by not covering the entire development life-cycles. Finding and selecting a suitable agent-oriented methodology can vary regarding the complexity of agent-based BI-systems. Hence, agent-based BI-systems are designed with different system specifications and choosing an appropriate agent-oriented methodology is a challenge for developers [14].

Literature highlights several efforts of researcher for selecting existing agent-oriented methodologies. In this regard, researchers propose multiple solutions in the format of evaluation frameworks, comparison methods, and approaches in [30–32]. Each of these evaluation processes are fulfilled based on different criteria and context to evaluate a limited number of methodologies. Still, none of these evaluation processes are known as a standard, or applicable method for comparing methodologies.

In this section for comprehending the analysis and design phases, we first briefly demonstrate our proposed new artifact BI-MAS life-cycle and applicable functional and non-functional requirements. Next, we demonstrate the selection processes of applicable agent-oriented methodology based on each phases of the standard development life-cycle (waterfall).

3.1 BI-MAS Life-Cycle

In this research, we consider those functional and non-functional requirements that are relevant for BI-systems together with DDM and MAS-technology. In general, the functional and none-functional requirements for a BI-system might be achieved from the business need within the context of organizational strategies, system structure and existing business processes. In this research, we assume to focus on the implementation of MAS technologies in the internal structure of BI-systems. Hence, agents play key roles in the BI-MAS development phases.

We explain the functional and non-functional requirements with representing a life-cycle for BI-MAS (as depicted in Fig. 1). The life-cycle starts with receiving an input from business stakeholders. The remaining life-cycle is defined as automated processes to find information about particular input-data. To implement agent-abilities for the entire BI-MS life-cycle, we assume that the workflow and data-flow are carried by MAS technologies, i.e., dispatching to data-sources, aggregation of information, data-mining processes, and so on. Table 1 illustrates all required notations used in BI-MAS life-cycle.

According to [14], system requirements must express the properties of a system and scenarios that specify the use-cases of intended systems and intended for implementation. For developing agent-based complex system, requirements and scenarios can be expressed in various degrees containing formal, semi-formal, and informal [14]. Therefore, we summarize to explain the functional-requirements of BI-MAS by considering MAS technology sequentially using sub-sections (A–F).

A. BI-MAS shall support a user interface (UI) for business users to navigate, explore and access into distributed data-sources, and receive information.
B. BI-MAS shall provide the means of aligning business intelligence, business process improvement and automation in internal logical work plan and data mining operation using MAS.
C. BI-MAS must contain those facilities, i.e., parallel ETL and OLAP processes, to speed-up data exploration processes while data collecting from different source systems into a more advanced discipline.
D. BI-MAS support parallel processes using MAS technology in data mining and knowledge discovery process that plays important roles in today's business area.

E. BI-MAS shall contain local data warehouse to store historical explored information within defined time specification.

F. BI-MAS shall emphasis on unifying data representation, cleaning, summarization, aggregation and understandable querying over transactional stores.

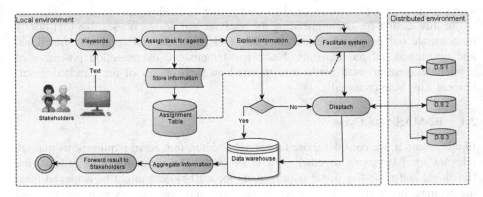

Fig. 1. BI-MAS life-cycle.

Table 1. Notation of BI-MAS life-cycle

	Starting point of the life-cycle
	Represents task that can be assigned to an agent
	Shows data storage servers
	Used to represent the condition
	Represents data warehouse
	Illustrates optional task
	Ending point of the life-cycle

Based on above mentioned assumption, we also deduce the following properties list as non-functional requirements that are applicable for BI-MAS.

Reactivity - defines ability of an agent to perceive and respond actively the environment in a timely manner.

Autonomy - represents ability of an agent to act independently without direct interaction of user.

Confidentiality - illustrates ability to protect the agents' data from other unauthorized agents, or other hosts.

Collaboration - indicates the ability of agents to interact with each other to achieve a common purpose, or objective.

Deliberately - shows activities of agents that represent an output in such a manner that output of on activity is input for next consecutive processes.

Accuracy - defines ability of agents to present the high quality of its performance during execution of several functions.

Reliability - illustrates the ability of agent during execution of several processes without any interruption whether errors occur in the system.

Trustability - represents ability in which an agent trusts another agent in same host to delegate part of their task in heterogeneous environments.

Security - indicates functions that are applied for agents to protect agent from another harmful agents, or hosts.

3.2   Agent-Oriented Methodologies

In this section, we briefly explain the nine well-known agent-oriented methodologies as follows. *Tropos* is an agent-oriented software engineering (AOSE) [15] methodology that covers software development processes in five phases of: early- requirements, late-requirements analysis, architectural-design, detailed design, and implementation. In early requirements analysis, Tropos focuses on the understanding of a problem by studying with organizational setting. Secondly, this methodology emphasizes on analysis phase for a deeper understanding of the environment where software must operate along with relevant functions and qualities. In the architectural-design phase, the system global architecture is defined in terms of sub-systems that are interconnected through data, control, and other dependencies. The required agents are specified at the micro-level and each agent's goals, roles and capabilities are specified in detail along with the interaction behaviors.

PASSI (Process for Agent Societies Specification and Implementation) [16] is a step-by-step requirement-to-code methodology for designing and developing multi-agent integrating system. The analysis and design phases of PASSI are determined and characterized by iterative step-by-step refinement. Therefore, producing a final stage to concrete design and implementation phases is based on the FIPA standard [16]. In addition, PASSI is composed of five models that address different design levels of abstraction such as system requirements-, agent society-, agent implementation-, code-, and deployment-models.

Prometheus [2] methodology is developed for building agent-based software systems. The main goal of Prometheus is to have a process with associated deliverables for industry practitioners and undergraduate students without a previous background in agents [17]. The Prometheus methodology consists of three phases such as system-specification, architectural-design, and detailed-design phases. The system-specification phase corresponds to the motivation layer and focuses on identifying the basic functionalities of a system. The architectural-design phase focuses on the types of functionalities that are delivered by agents. Additionally, this phase determines agent roles, agent-acquaintance diagrams, data type and protocols that are applicable in system architecture. Finally, the detailed-design phase looks at the internal characteristics of each agent and how it can fulfil its tasks within the overall system.

ADELFE [18] methodology developed to software engineering Adaptive-MAS (AMAS). In fact, adaptive software is used in situations where either an environment is unpredictable or a system is very open. This methodology guarantees that software is developed according to the AMAS theory to cover preliminary-requirements, final-requirements, analysis, design, implementation and tests. In the analysis phase, an engineer is guided to decide to use adaptive multi-agent technology and to identify an agent through the system and environment models. In the design phase, this methodology provides cooperative agent models and helps the developer to define local agent behavior.

MOBMAS (Methodology for Ontology-Based Multi-Agent Systems) [19] is a software engineering methodology that contains activities and associated steps to conduct the system development, techniques to assist a process, and a definition of models. The development process of MOBMAS is highly iterative and incremental between activities. In total, there are five activities, each focusing on a significant area of MAS development such as analysis-activity, MAS-organization design- activity, agent-internal design-activity, agent-interaction design-activity, architecture design-activity.

MaSE (Multi-agent Systems Engineering) [20] is an established object-oriented methodology that supports a complete life-cycle to design and develop agent-based systems. MaSE has been extended to an Organization-based MAS-Engineering (O-MaSE) framework. Additionally, O-MaSE is an architecture-independent methodology [2] that consists of three steps: capturing goals, applying use-cases, and refining roles. Consequently, the design phase has four steps: creating agent classes, constructing conversations, assembling agent classes, and system design. These steps are also called models that describe a process to guide a system developer from an initial system specification to system implementation. Furthermore, this methodology proposes nine classes of models under its life-cycle such as goal-hierarchy, use-case, sequence-diagrams, roles, concurrent-task, agent-classes, conversations, agent-architecture, and development-diagrams [20].

Gaia [38, 39], is known as one of the first complete methodologies for the analysis and design of MAS. This methodology is applied after gathering requirements that cover the analysis- and design phases. In the analysis phase, the role model and interaction model are constructed. The agent model, services model, and acquaintance model are constructed during detailed design stages. Additionally, the Gaia method has many similarities with MaSE [21]. In general, both MaSE and Gaia capture much of the same type of information from requirements. In Gaia methodology, most of the proposals concentrate on the analysis phase. As a MAS concept is quite complex, this methodology provides models and guidance that is near to some anthropomorphic modelling, which is convenient for understanding the system problem [21]. In Gaia, roles and services help to organize the functionality that is associated to an agent or a group of agents.

ROADMAP (Role-Oriented Analysis and Design for Multi-agent Programming) [17, 20] methodology extends Gaia with four improvements such as formal models of knowledge, role hierarchies, explicit representation of social structures, and incorporation of dynamic changes. In this methodology, a complex system is defined as a computational organization of interacting roles at the analysis stage optimized for quality goals, and populated with agents at the design stage. The roles in ROADMAP have runtime realization that allows runtime reasoning, social aspects modifying, and agent characterizing. In ROADMAP, the models that are constructed in analysis and design phases including use case-, environment-, knowledge-, role-(characterized by four attributes: responsibilities, permissions, activities and protocols), interaction-(contains protocol model), agent-, services-, acquaintance-models.

RAP (Radical Agent-oriented Process) [17, 40] is based on Agent-Object Relationship (AOR) modeling. The essential objective of RAP/AOR methodology is to enhance team productivity by agent-based work process management including both workflows and automatic interactions among team members in a system [2]. Unlike other mentioned agent-oriented methodologies, RAP/AOR is more concerned with distributed MAS. In AOR, several models are included, i.e., agents' actions, event perceptions, commitments, and claims. The Agent's role can be represented by AOR agent diagrams where different agent types may relate to each other through a relationship of generalization and aggregation.

3.3 The Evaluation Results

Based on the defined life-cycle for BI-MAS (discussed in Sect. 3.1), we need to find a fitting software engineering development method out of these nine well-known agent-oriented methodologies. Each of these methodologies has its own respective concept, modeling language, processes, specifications, principles, etc. It is very difficult to select one of them by chance without either understanding the development phases, or having an assessment results. Since the selection process of an agent-oriented methodology is a challenge [30–32]. finding commonalities between these proposed evaluation frameworks for performing the evolution process, must be well specified.

Table 2. Notations for evaluation process of agent-oriented methodologies.

Notations	Descriptions
F	For fully coverage
M	For mostly coverage
P	For partial coverage
N	For none or zero coverage
U	The sum of total calculation for each methodology
m	The number of agent-oriented methodologies
β	Represents the respective weights

On the basis of our study, each of these discussed methodologies has relevant development steps in their life-cycle similar to the waterfall model [22]. To evaluate these methodologies, we need criteria in our evaluation processes that fall into the

Software Requirement categories termed *Functional* and *Non-functional* (as illustrated in Table 3). Thus, the evaluation procedure with detailed description requires an equation that takes into consideration each phase of development life-cycle [23]. Thus, we define Eq. 1, in which several criteria are required as Table 2 illustrates.

As outlined in Table 3, to achieve the utility U of an agent-oriented methodology m, for computing processes, each of these defined variables {F, M, P and N} receives a score on the scale {3 | 2 | 1 | 0}. It means that 3 denotes full coverage, 2 mostly coverage, 1 partial coverage, and 0 for none coverage of each components that is used in each phases of a software development life-cycle. The respective weights (βs), for the requirements for fully F, the mostly M, and partially P, can be set in many ways. Our main concern from respective weights β in this equation is $\beta_F > \beta_M > \beta_P > \beta_N$.

$$U^m = \beta_F F^m + \beta_m \sum_i M_i^m + \beta_P \sum_j P_j^m + \beta_N \sum_k N_k^m \tag{1}$$

Furthermore, the evaluation results of Eq. 1 are outlined in Table 3. As a result, if we consider the second row (*Non-functional*) result that indicates very low scores of these respective nine methodologies, certain methodologies do not cover non-functional requirements at all, while others have merely low scores. Similarly, when we consider *Testing*, again the coverage is either none, or very low score. We conclude that none of these listed agent-oriented methodologies support fully the BI-MAS development life-cycle individually and none of these methodologies has high scores from initial-stage via very advanced-level of implementation to test processes.

Table 3. Evaluation result of agent-oriented methodologies.

Development life-cycle	Tropos	PASSI	Prometheus	ADELFE	MOBMAS	MaSE	Gaia	ROADMAP	RAP/AOR
Functional	3	2	3	2	2	2	2	3	2
Non-functional	2	0	0	0	0	1	0	2	2
Analysis	3	1	1	3	2	3	3	3	3
Design	2	2	2	2	2	3	2	3	3
Implementation	2	1	1	2	1	1	1	3	3
Testing (V&V)	1	0	0	0	1	2	1	1	1
Utility m	12	6	7	9	8	12	9	15	14

On the other hand, our studies discover that the Gaia methodology, the extended agent-oriented methodologies ROADMAP and RAP/AOR score best, especially during the analysis and design phases (also shown in Fig. 2). The MaSE methodology also has good scores in the analysis and design phases while the *Non-functional* and *Testing* phases have very low scores. Additionally, the ROADMAP and RAP/AOR have a comparable foundation for their development life-cycle and support each other. It means both offer promising options in the analysis- and design-phases during deployment [2].

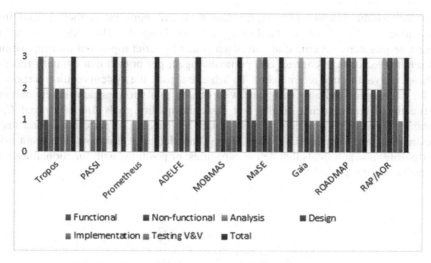

Fig. 2. Statistical analysis of agent-oriented methodologies

According to [17, 43], ROADMAP and RAP/AOR provide a common framework of system features for specifying, designing, developing, and implementing intelligent agent systems. At the level of computational design and implementation of these methodologies the focus rests on different kinds of models from various aspects. For instance, a goal-model to define actors in the intended system, a domain-model to identify related objects in the system domain, a knowledge-model to indicate the properties of objects in their respective contexts, an interaction-model to represent MAS realistic interactions, and behavior-model to address the decision making and performing activities of each agent. These contexts support us to select these two methodologies to cover the analysis- and design-phases for BI-MAS life-cycle.

4   Detailed Design Phase of BI-MAS Architecture

To pursue the ROADMAP and RAP/AOR methodologies along with AOM techniques [2] for fulfilling the analysis- and design-phases, our objective is to develop several required models that transfer the defined functional and none-functional requirements along with the terms of agent functions, roles, and behaviors. To provide a clearer understanding from the analysis- and design phases, we consider the graphical notation that is sufficiently expressed to handle the complexity of BI-MAS in following subsections.

4.1   The Goal Model

In the goal model of BI-MAS that is depicted in Fig. 3, we first present the root-functional goal of *Run BI-system* with the attached role of *Stakeholder*. According to ROADMAP and RAP/AOR methodologies [2], the root-functional goal is called the value proposition that is too complex and therefore must be further refined into manageable functional sub-goals. During characterizing the BI-MAS goal model, simple tree-hierarchy schema diagram is generated by strict top-down decomposition. This schema leads us to connect one particular agent per branch that contains relative sub-tasks relevant to one requirement. To achieve a goal, the system requires a specific role for each agent and also sub-goals with quality goals to represent functional and non-functional requirements. In the first case, the quality goals *Autonomy* and *Collaboration* mean that the agents of a BI-system are capable of performing their tasks autonomously and support each other during knowledge exploration. The main goal includes roles and sub-goals that define capacities, or positions with functionalities that are needed for the BI-MAS.

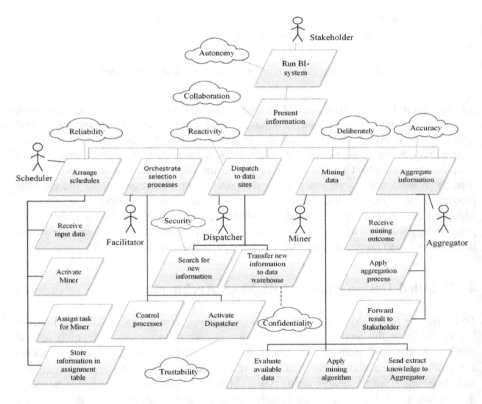

Fig. 3. The goal model of the BI-MAS.

We decompose the main goal that is associated with *Present information* into smaller related sub-goals such as *Arrange schedules, Orchestrate selection processes, Dispatch to data-sites, Mining data,* and *Aggregate information.* The *Arrange schedules* goal is decomposed into four sub-goals of *Receive input data, Activate Miner,*

Assign task for Miner, and *Store information in assignment table.* Additionally, this goal is attached to the role of *Scheduler* and the quality goal of *Reliability* that represents the responsibility of an agent for setting an assignment to other single, or a group of agents based on the received input data. The *Orchestrate selection processes* goal includes two sub-goals *Control processes* and *Activate Dispatcher* with a quality goal of *Trustability.*

Furthermore, this goal is attached to the role of *Facilitator* that is responsible for activation and termination of *Dispatcher* agents. The *Dispatch to data sites* goal comprises also two sub-goals *Search for new information* with the quality goal of *Security,* and *Transfer new information to data warehouse* with an attached quality goal of *Confidentiality.* We attach the role of *Dispatcher* to this goal that is responsible to explore new information from different data-sites for transferal to a data warehouse. The *Mining data* goal contains three sub-goals *Evaluate available data, Apply mining algorithm,* and *Send extract knowledge to Aggregator.* This goal also attaches to the role *Miner* together with the quality goal of *Deliberately* that represents the performance of agents during knowledge exploration for sharing with other mining processes. The *Aggregate information* goal includes three sub-goals *Receive mining outcome, Apply aggregation process,* and *Forward result to Stakeholder.* We describe these goals with the attached role of *Aggregator* with the quality goal of *Accuracy* that is responsible to obtain knowledge from other miner agent/agents separately and after the modification and collection processes, it submits the result to the *Stakeholder.*

4.2   The Domain Model

With respect to ROADMAP and RAP/AOR methodologies [2], for each defined role there must be an agent mapped in. The model that shows knowledge about the environments and illustrates relationships of agents is called domain model [24]. In this section, we discuss the domain model that represents the entities of the problem domain that are relevant for BI-MAS environments (shown in Fig. 4). This model describes the main domain entities, the agents' roles, and their relationships with each other within two environments. In fact, an agent environment produces and stores objects that can be modeled as resources, which are accessed by agents [2]. In this regard, we consider two types of environments in which the agents either exist or migrate to. The local environment where all the activities of agents perform between each other, are also called agent host [6]. The distributed environment where the *Dispatcher* agents can migrate to is related to *Data sites of system.* The agent that plays the role of *Stakeholder* can interact with real BI users in a local environment via *User interface.*

All other remaining agents are software agents identifiable based on their activities between these two environments. For instance, the *Scheduler* agent is responsible for activating and assigning tasks to the *Miner* that is situated in the local environment. The *System assignment table* comprises domain entities where all information about agents and their responsibilities are stored that belong to this environment. The *Miner* agent is responsible for discovering knowledge from a *Data warehouse* that is modeled as a

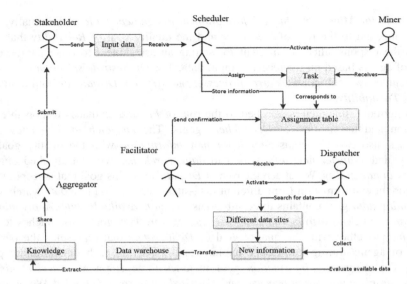

Fig. 4. The domain model of the BI-MAS.

domain entity of the local environment. In addition, the *Aggregator* has a responsibility to summarize the collected information from the *Miner* agent and submit to *Stakeholder*. Furthermore, the data sites of a system is modeled as a domain entity (*Different data sites*) in the domain model that belongs to a distributed environment. Here, the *Dispatcher* agent is responsible for transferring new information from a distributed to local environment.

In the next section, we derive a knowledge model from the domain entities that is relevant for the agent knowledge base.

4.3   The Knowledge Model

A knowledge model can be viewed as an ontology that provides a framework of knowledge for agents of problem domains [2]. The agents' knowledge model demonstrates the agent role, internal knowledge and the relationship with objects in the environment. Agents represent information about itself via knowledge attributes that are intrinsic properties of an agent. There are two kinds of attributes: numeric- and non-numeric attributes. A non-numeric attribute of an agent represents one or more quality dimensions. The knowledge-attribute types are string, integer, real, boolean, date and enumeration. In the knowledge model depicted in Fig. 5, each agent is modeled by representing a type, name, id, and relationships within an environment. For instance, the *Miner* agent knows about the task that is assigned by the *Scheduler* agent, and knows about the *DataWareHouse* where it can search to explore new knowledge.

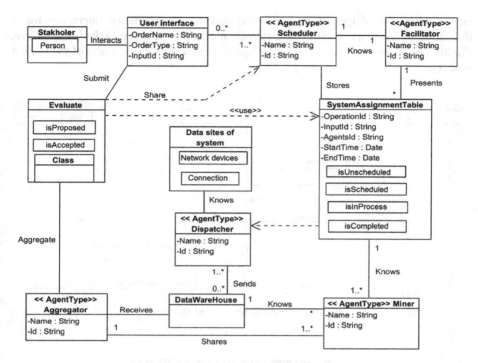

Fig. 5. The knowledge model of the BI-MAS.

Several objects of various types such as *SystemAssignmentTable, DataWareHouse, Data sites of system,* and *User interface,* are defined in the knowledge model. These objects are shared between all agents playing different roles in the BI-MAS component. For instance, the *SystemAssignmentTable* in which all information about the agents activities are stored by the *Scheduler* agent is shared between the *Facilitator, Miner,* and *Dispatcher* agents. Moreover, in each object of the knowledge model, several related attributes and predicates are defined. For example, the object of *SystemAssignmentTable* in Fig. 5 describes the attributes of *OperationId, InputId, AgentId, StartTime,* and *EndTime* that represent the information relevant to each agent. Furthermore, *SystemAssignmentTable* illustrates several status predicates of agents such as *isUnscheduled, isScheduled, isInProcess,* and *isCompleted* that are demonstrates the status of agents. Next, we describe more about what message flow and interaction occur between agents involved in BI-MAS.

4.4   The Interaction Model

According to ROADMAP and RAP/AOR methodologies [2], the interaction modeling must represents the interaction links between multiple agents of a system. Through interaction modeling, we exhibit a clear concept for an observer to understand what message flow and interaction occurs between agents and how the sequence of actions are performed by each agent. Additionally, the interaction model can be captured by any of interaction-diagrams, or interaction-sequence diagrams, or interaction-frame diagrams. Below, we depict the interaction-diagram between *Scheduler*, *Miner*, *Facilitator*, *Dispatcher*, and *Aggregator* agents with a depicted scenario of the mining processes.

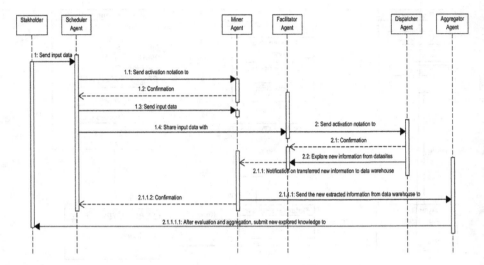

Fig. 6. The interaction model of the BI-MAS.

In interaction model depicted in Fig. 6, the arrows between agents demonstrate the existence of an interaction link that allowing an agent to initiate interaction with another agent. In general, an interaction can occur either by sending a message to another agent or performing a physical action affecting to it. Information about each interaction is extracted from responsibility, or role of each agent. In this diagram, each action event is characterized by a sequence number. These numbers constitute an interaction sequence between agents that are involved in this diagram.

Next, we elaborate the system scenario and behavior model that identify the sequence of various activities in which each agent plays a specific role in BI-MAS concept.

4.5   The Behavior Model

A behavior model is a scenario that is described to achieve the system goal by system agents [2]. A scenario can be defined also as collective activities that involve either a single, or multiple agents. Similarly, a scenario is illustrated with sub-scenarios that are corresponding to sub-goals of the system. In this section, we present a motivational scenario for each agent that has a specific role in BI-MAS. This scenario is based on the format of a goal-based use-case that is originally used in the RAP/AOR methodology [25]. For instance, the scenario corresponding to the goal *"Present Information"* (as illustrated in Table 4) has five sub-scenarios with respective sub-goals such as *"Arrange schedules"*, *"Orchestrate selection processes"*, *"Dispatch to data sites"*, *"Mining data"*, and *"Aggregate information"*. In addition, during the behavior modeling, a scenario can be triggered by a situation that involves the agent initiating related scenarios. For example, a scenario to reach for system goal *"Present Information"* is triggered by an action *"Send input data"* that is performed by a *Stakeholder* (shown in Table 4).

Table 4. A Scenario for achieving the goal "Present Information".

SCENARIO 1					
Goal	Present information				
Initiator	Stakeholder				
Trigger	Send input data by Stakeholder				
Description					
Condition	Step	Activity	Agent type/roles	Resources	Quality goals
	1	Arrange Schedules (Scenario 2)	Scheduler	Input data	Reliability
If data warehouse is empty	2	Orchestrate selection processes (Scenario 3)	Facilitator		Trustability
	3	Dispatch to data-sites (Scenario 4)	Dispatcher	Input data	Reactivity, security, and confidentiality
	4	Mining data (Scenario 5)	Miner	Data warehouse	Deliberately
If new explored information is not same	5	Aggregate information (Scenario 6)	Aggregator	New knowledge	Accuracy

Table 5. A Scenario for achieving the goal "Arrange schedules".

SCENARIO 2					
Goal	Arrange schedules				
Initiator	Scheduler				
Trigger	Input data received by Scheduler				
Description					
Condition	Step	Activity	Agent type/roles	Resources	Reliability
	1	Receive input data	Scheduler	Input data	
If more than one input data arrive	2	Activate Miner agent	Scheduler	Input data	
	3	Assign task for Miner	Scheduler	Input data	
	4	Store information in assignment table	Scheduler	Input data	

Table 6. A Scenario for achieving the goal "Orchestrate selection processes".

SCENARIO 3					
Goal	Orchestrate selection processes				
Initiator	Facilitator				
Trigger	Assign task for Miner by Scheduler				
Description					
Condition	Step	Activity	Agent type/roles	Resources	Quality goals
	1	Control processes	Facilitator	Assignment table	Trustability
If the requested data more than one	2	Activate Dispatcher	Facilitator		

Table 7. A Scenario for achieving the goal "Dispatch to data sites".

SCENARIO 4					
Goal	Dispatch to data sites				
Initiator	Dispatcher				
Trigger	Agent function activates by Facilitator				
Description					
Condition	Step	Activity	Agent type/roles	Resources	Quality goals
If data belong to different data sites	1	Dispatch to data-sites	Dispatcher	Different data sites	Reactivity
If the new information is not already in data warehouse	2	Search for new information	Dispatcher	Data warehouse	Security
	3	Transfer new information to data warehouse	Dispatcher	Data warehouse	Confidentiality

Table 8. A Scenario for achieving the goal "Mining data".

SCENARIO 5					
Goal	Mining data				
Initiator	Miner				
Trigger	New information transferred to data warehouse by Dispatcher				
Description					
Condition	Step	Activity	Agent type/roles	Resources	Quality goals
	1	Evaluate available data	Miner	Data warehouse	Deliberately
If data transferred more than one data site	2	Apply mining algorithm	Miner		
	3	Send extract knowledge to Aggregator	Miner		

Table 9. A Scenario for achieving the goal "Aggregate information".

SCENARIO 6					
Goal	Aggregate information				
Initiator	Aggregator				
Trigger	New knowledge shared by Aggregator				
Description					
Condition	Step	Activity	Agent type/roles	Resources	Quality goals
	1	Receive mining outcome	Aggregator	New knowledge	Accuracy
	2	Apply aggregation process	Aggregator	New knowledge	
	3	Forward result to Stakeholders	Aggregator	New knowledge	

Agent behavior models are platform-independent [2], and can be expressed only in terms of an abstract agent architecture. According to the abstract agent, an agent behavior is determined by a controller based on agent perceptions and knowledge. In this model (shown in Fig. 7), the controller is modeled in the terms of roles and behavior that is relevant for BI-MAS requirements. In fact, the agent's role starts a sequence of activities comprising various actions. During the execution of each cycle of an abstract agent, each agent role can be recognized and triggered by a perception. Each defined agent can percept a message, or action that is performed by other agents. In addition, some types of roles are not triggered in a system and occur as a start event once per each execution cycle of an abstract agent. For instance, Fig. 7 represents the behavior of an agent called *Scheduler* who initiates the execution life-cycle process by perceiving input data.

The sequences of roles that are marked with R mean the following. R1 shows activities "*Receive input data*", "*Activate Miner agent*", "*Assign task for Miner*", and "*Store information in assignment table*" that are listed in Table 5. The condition attached to role R3 means it is triggered only if the requested input data does not exist in the *data warehouse*. According to R3, the *Facilitator* performs activities that are outlined in Table 6. Consequently, role R4 follows the *Dispatcher* with performing activity types "*Search for data*" and "*Transfer new information to data warehouse*" (as illustrated in Table 7). When the data is transferred in the *data warehouse*, the role R2 starts the activities of type "*Evaluate available data*", "*Apply mining algorithm*", and "*Send extract knowledge to Aggregator*" (as illustrated in Table 8). Finally, after receiving a confirmation message about new explored information, the role R5 is started with activities types "*Receive mining outcome*", "*Apply aggregation process*", and "*Forward result to Stakeholders*" by *Aggregator* (as illustrated in Table 9). To know more about the notations that are used in agent behavior model demonstrated in Fig. 7, we refer the reader to reference [2].

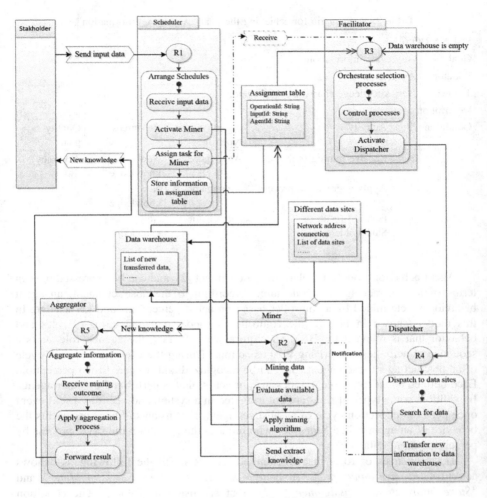

Fig. 7. The behavior model of the BI-MAS.

4.6 Overview of BI-MAS

As illustrated in Fig. 8, the baseline system requirements are transferred into a high-level overview of BI-MAS by performing the analysis- and design-phases with the support of the ROADMAP and RAP/AOR methodologies.

In Fig. 8, we present a three layered view for the BI-MAS framework. In this research, the main focus is on the agent-level and therefore, the important components of the BI-MAS comprise agents with different roles defined into an integrated layer-based structure. Each layer comprises a single, or multiple agents that have key roles to perform specific functional requirement (discussed in Sect. 3.1). To determine the correlation among each layer, we assume to have one agent in the interface level and the remaining must be defined on the operating level. For instance, the *Stakeholder*

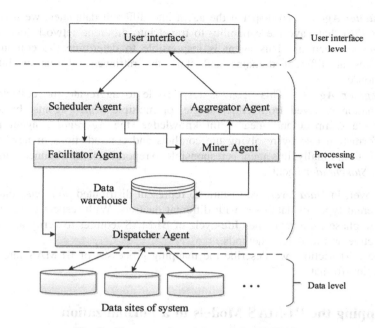

Fig. 8. The general overview of the BI-MAS.

agent is defined as a human agent that interacts with the BI-MAS stakeholders through *User Interface*. For simplicity, the *User Interface* module comprises functionality to capture the input data (keywords) and reports back the research result (new explored information) to stakeholders via BI-MAS system interface. It means that the operation of BI-MAS follows sequences, i.e., an operation starts from the top (*User Interface*) to bottom, and from the bottom to the top, while the research results are found and transferred to stakeholders. The remaining agents (as shown in Fig. 8) are defined in the *processing level* as follows:

Scheduler Agent - agent is responsible to This agent is responsible to receive the requested keywords and determine the types of operations defined under the BI-system and creates a work plan for other agents accordingly. After assigning tasks to a single-, or group of agents, updated information is stored into the *System Assignment Table* (as shown in Fig. 4).

Facilitator Agent - This agent is responsible to facilitate the mining process due to activation and termination of the *Dispatcher* agents. Moreover, this agent comprises a knowledge module that stores the history of requested keywords and previously retrieved information in the data warehouse that helps the *Miner* agent to explore new information without -waiting for the *Dispatcher* agent.

Miner Agent - The *Miner* agent plays an important role in the mining of data from the local environment (*Data warehouse*) by deploying mining algorithms. Additionally, this agent comprises a module to share the explored information automatically with other *Aggregator* agent.

Dispatcher Agent - To dispatch the agent into different data-sites, we use mobile agents [26] that have the capability to travel into different network locations via Internet connections. This agent is responsible to determine the computational resources at different domains and store the retrieved information into data warehouse.

Aggregator Agent - This agent is responsible to aggregate the collected new information received from either single, or multiple *Miner* agents. In order to present a compact and meaningful knowledge, the *Aggregator* agent plays a transformation role by resolving the conflicts and contradictions of newly mined information. Finally, this agent is responsible to report back the obtained knowledge to the *Stakeholder* agent.

Moreover, in *Data level*, we assume to represent distributed *data sites* that might have different types of databases with different datasets. With respect to the analysis and design phases, the developed life-cycle of BI-MAS requires to verify and validate with effective and technical methods.

In the next section, we describe the mapping processes of BI-MAS into several formalization format.

5   Mapping the BI-MAS Models to a Formalization

In general, the verification and validation (V&V) [27] processes are conducted to assure the quality of product-, or development life-cycle based on system requirements. According to [28], the construction of V&V processes of self-adaptive software systems such as agent-based distributed systems have remained a very challenging task for developers. There is a need for novel V&V methods to provide assurance of the result for the entire life-cycle of complex-systems. As illustrated in Table 3, even the V&V processes are not supported fully by any of existing agent-oriented methodologies. In this section, we intend to represent a new approach of V&V processes for BI-MAS with new methods and tools.

In this regard, our studies show that CPN-tool receive interest of researchers for designing V&V processes of distributed systems [29]. With respect to [47, 48], Colored Petri Nets (CPN) is a notation for the modeling and validating of systems in which concurrency, communication, and synchronization are the foci. In order to formulate the BI-MAS life-cycle, it is important to map BI-system models to a formal and deterministic notation that allow us to fulfill the V&V processes. To accomplish the V&V processes, we consider CPN-tools that supports extensions with time, color, and hierarchy for modeling and analysis of distributed systems by a graphical simulation tool [30]. The CPN language allows to organize a model as a set of modules, and it includes a time concept for representing the time token to execute events in the modelled system. The modules connect with each other through a set of well-defined interfaces in a similar way as known from many modern programming languages.

Table 10. Acronyms, names and descriptions of token colors.

Level	CPN module	Data property	Description	Type
1	BI-MAS	sc	Scheduler unique id based on that receives input data from stakeholder/stakeholders	Integer
		sh	Based on this numbers, several stakeholders can request multiple keywords with specific id	
		mid	Activating numbers for Miner agents	
		key	Stakeholder keywords searching for new knowledge	String
		s	The final explored new knowledge	
		t	The time sequences in which the data is arrived in a place from different data-sites	Time
2	Activate Agents	sh1	The stakeholder id that is stored in temporary place to fulfill the aggregation processes	Integer
		key1	The stakeholder keyword that is stored in temporary place to fulfill the aggregation processes	String
		s1	The final explored knowledge that is stored in temporary place to fulfill the aggregation processes	
		t1	The time sequences that are used to compare two results arrived from different data-sites based on one keyword	Time
3	Search for data-sites	dsid	Activated numbers for Dispatcher agents	Integer
		t	The time that is generated for each sequence of data, which is explored from different data-sites	Time
4	Data-site1	sid1	Unique id that is related to data-site1	String
		f1_key	Finding key based on input keyword on data-site1	
		s_r1	Present the result for searching keywords on data-site1	
	Data-site2	sid2	Unique id that is related to data-site2	String
		f2_key	Finding key based on input keyword on data-site2	
		s_r2	Present the result for searching keywords on data-site2	
	Data-site3	sid3	Unique id that is related to data-site3	String
		f3_key	Finding key based on input keyword on data-site3	
		s_r3	Present the result for searching keywords on data-site3	

The CPN model contains places, drawn as ellipses or circles, transitions drawn as rectangular boxes, a number of directed arrows connecting places and transitions, and finally some textual inscriptions. For instance, places and transitions are called nodes that are connected with directed arrows. An arrow always connects a place to a transition, or a transition to a place. In a CPN model, places may hold multiple tokens that carry color (i.e., attributes with value). In addition, transitions are ready to fire when all input places hold the required sets of tokens and produce condition-adhering tokens into output places. In this section, we present step by step processes of mapping and formal transformation of BI-MAS life-cycle that is prerequisite for V&V processes.

5.1   Related BI-MAS Data

For the formalization processes, the data elements of the BI-MAS model and sub-modules are declared for 4 refinement hierarchical-levels. Table 10 lists all the relevant token colors with their hierarchic refinement availability that is used for all lower- but not for any higher hierarchy levels. In the left column of Table 10, number 1–4 represents sequentially from level 1 to level 4 the lowest refinement levels of the BI-MAS components. In the second- and third columns showing are the name of nested modules and token colors that are used during the formalization of BI-MAS models. The fourth column textually explains the data-flow properties of the BI-MAS life-cycle. Finally, the fifth column presents the token colors properties while their types are defined either integer, string, or time. The integer-type of tokens is used as identification number and string-type tokens can be either stakeholder input keywords, or a matching result for corresponding search key. Time-type tokens also called time stamps that can be used for different purposes in CPN models. In this paper, we use time as a sequence number associated with objects to specify the first and last arrival of tokens from other places into targeted places.

5.2   Formalized BI-MAS with Nested-Modules

With respect to [31], to resolve the complexity of a distributed system, the design processes must be produced by modularity, regularity, and hierarchy characteristics. As CPN-tools support hierarchy nested-modules, we use this property of CPN for applying nested modules to cover the entire BI-MAS life-cycle. The top-level module of BI-MAS depicted on Fig. 9 formalizes the cooperative environment of BI-MAS that are used for data exploration, or data mining in distributed systems. Here, we assume that each token represents several unique ids associated with search keys and matching search results. This associated id helps the *Aggregator* agent to prevent conflicts and contradictions at the end of life-cycle. The searching- and mining processes in this complex system are well formed based on discrete business-process specifications that start with a unique state, in which the tasks are processed in a parallel structure by agents that lead to a unique end state. In this figure the life-cycle starts the processes either by receiving a single *input data* or multiple *input data* (as discussed in Table 11) simultaneously as requested by *Stakeholders*. The mining processes ends while agents find new information based on the *input data* from different *data-sites* (shown in Fig. 12).

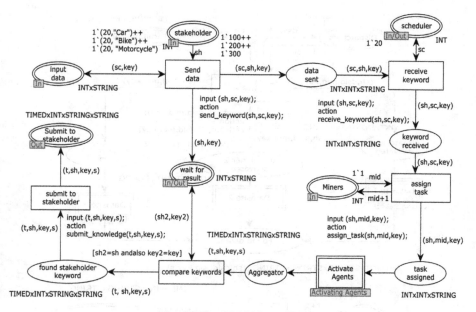

Fig. 9. The BI-MAS top level.

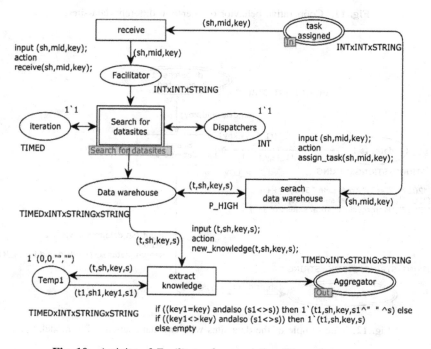

Fig. 10. Activity of *Facilitator* for generating *Dispatcher* agents.

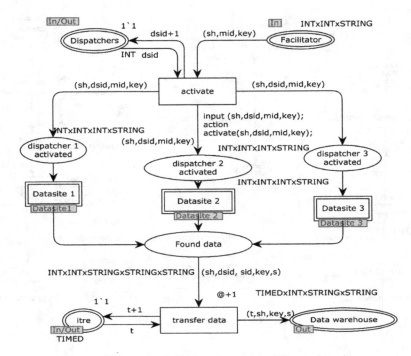

Fig. 11. Cooperative behavior of agents in different data-sites.

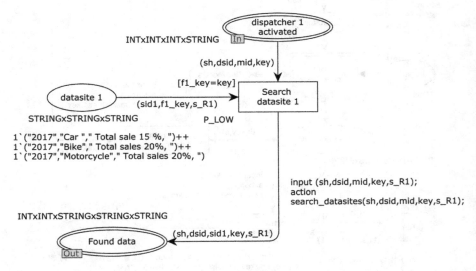

Fig. 12. One sample of the data-sites with two data sets in CPN model.

The second model covers the formalization of the two main agents' life-cycles relevant to *Facilitator* and *Aggregator*, as shown with double lined rectangles (*Activate Agents*) in Fig. 9. As demonstrated in Fig. 10, the life-cycles perform sub-module looking to *data-sites* in automated structure, while the *Facilitator* receives at least one keyword, or multiple keywords simultaneously. Here, for each search-key multiple *Dispatchers* can be activated based on the number of *data-sites*. In this part, the agents perform parallel computing that are used for quick access and manipulation of such distributed *data-sites*. In addition, the activity of *Aggregator* proceeds by a normalization function that is necessary for avoiding conflicts and contradicting data sets. For instance, here we define an evaluation function that is applied for outcome results of matching search-keys to normalize the consecutive search results depending on one keyword (explained in Table 11).

The construction of the third model is related to double lined rectangles (*Search for data-sites*) in Fig. 10. As illustrated in Fig. 11, the activities of an agent depend on the cooperative behavior of agents to exploit such computing environments for scaling up the data mining process. Here, the module shows that the data-mining processes can fulfill without loading all data sets into a single site. Instead, the resulting mining process transfers data into the warehouse. For V&V processes of BI-MAS, we assume to have three data-sites with different data-sets. Due to page limitation, we present here one sample, i.e., data-site 1 in Fig. 12. The remaining two other data-sites have the same structure while only the contents are different.

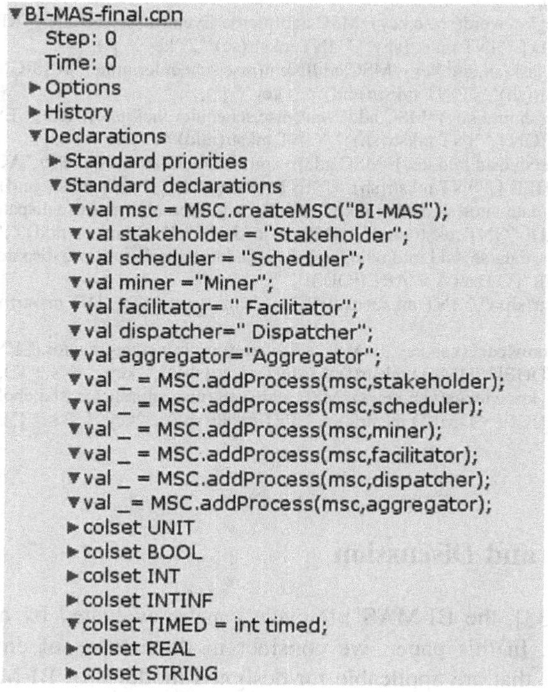

```
▼BI-MAS-final.cpn
    Step: 0
    Time: 0
  ▶ Options
  ▶ History
  ▼ Declarations
    ▶ Standard priorities
    ▼ Standard declarations
      ▼val msc = MSC.createMSC("BI-MAS");
      ▼val stakeholder = "Stakeholder";
      ▼val scheduler = "Scheduler";
      ▼val miner ="Miner";
      ▼val facilitator= " Facilitator";
      ▼val dispatcher=" Dispatcher";
      ▼val aggregator="Aggregator";
      ▼val _ = MSC.addProcess(msc,stakeholder);
      ▼val _ = MSC.addProcess(msc,scheduler);
      ▼val _ = MSC.addProcess(msc,miner);
      ▼val _ = MSC.addProcess(msc,facilitator);
      ▼val _ = MSC.addProcess(msc,dispatcher);
      ▼val _= MSC.addProcess(msc,aggregator);
      ▶ colset UNIT
      ▶ colset BOOL
      ▶ colset INT
      ▶ colset INTINF
      ▼ colset TIMED = int timed;
      ▶ colset REAL
      ▶ colset STRING
```

Fig. 13. Agent-interaction model with built-in functions in CPN-Tools

5.3   Transformations of BI-MAS Models to CPN Tools

The transformation of the corresponding types of conceptual models (explained in Sects. 4.1, 4.2, 4.3, 4.4, 4.5, 4.6), require syntactically procedures to represent the automated simulation results using CPN-tools. The mapping constructs of knowledge-, behavior- and interaction attributes can be transferred either by built-in functions, or user defined properties of CPN-tools. Message Sequence Charts (MSC) is well-defined functions for system engineering are used to present the communication messages of sender and receiver in complex systems [32]. In addition, MSC functions are used for adding new processes, presenting events between processes, and adding internal events into single process, or external events between two objects.

Figure 13 shows the defined functions of transforming such as knowledge-, interaction- and behavior models of BI-MAS using CPN-tools. To setup the MSC function, it is required to add declarations to CPN. We create one process for each agent as shown in Fig. 13, i.e. *Stakeholder, Scheduler, Miner, Facilitator, Dispatcher* and *Aggregator*. Based on scenarios of behavior models in Sect. 4.5, it is important to declare pockets for a sender who can send data from one place to receivers. In List.1, we explain the compound data types that are transmitted between agents by using MSC function.

List1.	*MSC functions to describe the pockets contain*
1	fun send_keyword(sh,sc,key)=MSC.addEvent(msc,stakeholder,scheduler, "SEARCH ["^key^"]");
2	fun receive_keyword(sh,sc,key)=MSC.addInternalEvent(msc,scheduler, "RECEIVE KEYWORD ["^INT.mkstr(sh)^","^INT.mkstr(sc)^","^key^"]");
3	fun assign_task(sh,mid,key)=MSC.addEvent(msc,scheduler,miner,"ASSIGN TASK ["^INT.mkstr(sh)^","^INT.mkstr(mid)^","^key^"]");
4	fun receive(sh,mid,key)=MSC.addEvent(msc,scheduler,facilitator,"REQUEST DISPATCHER ACTIVATION ["^INT.mkstr(sh)^","^INT.mkstr(mid)^","^key^"]");
5	fun activate(sh,dsid,mid,key)=MSC.addEvent(msc,facilitator,dispatcher,"ACTIVATE DISPATCHER ["^INT.mkstr(sh)^","^INT.mkstr(dsid)^","^INT.mkstr(mid)^","^key^"]");
6	fun search_data-sites(sh,dsid,mid,key,s)=MSC.addInternalEvent(msc,dispatcher, " FOUND KEYWORD["^INT.mkstr(sh)^","^INT.mkstr(dsid)^","^INT.mkstr(mid)^","^key^","^s^"]");
7	fun transfer_data(sh,dsid,mid,ndid,key)=MSC.addInternalEvent(msc,dispatcher, " TRANSFER TO DATA WAREHOUSE["^INT.mkstr(sh)^","^INT.mkstr(dsid)^","^INT.mkstr(mid)^","^INT.mkstr(ndid)^","^key^"]");
8	fun new_knowledge(t,sh,key,s)=MSC.addEvent(msc,miner,aggregator, " EXTRACT KNOWLEDGE["^TIMED.mkstr(t)^","^INT.mkstr(sh)^","^key^","^s^"]");
9	fun submit_knowledge(t,sh,key,s)=MSC.addEvent(msc,aggregator, stakeholder, " SUBMIT KNOWLEDGE["^TIMED.mkstr(t)^","^INT.mkstr(sh)^","^key^","^s^"]");

6   Evaluation and Discussion

With respect to [33], the BI-MAS life-cycle can be evaluated by applying different types of methods. In this paper, we consider in three types of empirical and non-empirical methods that are applicable for designed modules of BI-MAS as follows.

6.1   Validation and Verification of BI-MAS

As CPN models are executable and are used to model and specify the behavior of agents in BI-MAS, this section presents the visualization and simulation result in CPN models. For V&V processes, we present the simulation results that are generated automatically. Each module of CPN can be simulated interactively, or automatically. For interactive simulation, we use Message Sequence Chart (MSC) [32] that generates automated results similar to single-step debugging. It also provides a way to 'walk through' into CPN models that investigate different scenarios in detail and check whether the model performs as expected. For simulating processes, we use two different scenarios as illustrated in Table 11 comprising testing and performance analysis.

Table 11. Scenario for simulation process of BI-MAS.

NO.	Scenario description	Initial input	Final output	Test-goal
I	Single input scenario- In this scenario, we assume that an organization has business for three products such as *Car, Bike*, and *Motorcycle*. Based on the number of products, three data-sites located in different physical locations containing different information. In this scenario, the stakeholder requests to receive particular information of type (Car) product	Searching for ("Car")	As shown in Fig. 14, the output is aggregated information from different data-sites only regarding Car	To test the workflow and data-flow for single- input as a token for the entire life-cycle of BI-MAS
II	Multiple-input scenario- The stored information is the same as in the previous scenario. Based on the number of products, three data-sites contain different information. In this scenario, the takeholder requests information regarding multiple inputs simultaneously	Searching for ("Car", "Bike", "Motor cycle") at same time	Due to page limitation, the outputs figures can not be demonstrated	To test the workflow and data-flow for parallel processing the life-cycle of BI-MAS

Figure 14 shows an example of an MSC result during the execution of Scenario-I. The MSC has six columns for this behavior- and interaction scenario, i.e. explained in Sect. 4.5. The leftmost column represent the senders and the rightmost columns represent the receivers. The MSC captures a scenario where the first data packet sent by the *Stakeholder* and the four middle columns represent the sender and receiver of the BI-MAS life-cycle. Finally, this process is ended where the *Aggregator* transmits the data packet (e.g., contains explored new information) to *Stakeholder*.

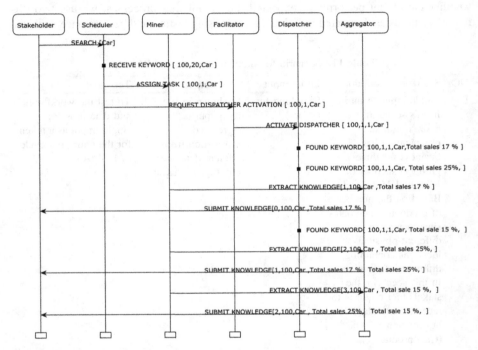

Fig. 14. MSC generated result for the BI-MAS life-cycle using CPN-tools

6.2   BI-MAS Models Properties

Colored Petri Nets is a formal modeling language that is well suited for modeling, validating and analyzing larger and complex systems. CPN Tools supports state spaces for hierarchical networks and offers facilities for collecting data during simulations and for generating different kinds of performance analysis reports. With respect to [48–53], the state space calculation and analysis considers each node that is involved in graphical representation of CPN models. Therefore, for testing and performance analysis of the BI-MAS life-cycle, we select the standard state space analysis instruments to collect data about the system performance.

In the first step, we consider the state space generation report for the BI-MAS life-cycle. The statistic result of state space report outlined in Table 12 provides basic information about size of behavioral properties and involved nodes. The first part of this result is generated based on single input Scenario-I (i.e., listed in Table 11) and the remaining parts of table are related to Scenario-II where we assume that 3 *Stakeholders* searching for 3 different keywords. In the fourth column of the table, one sample of dead transition is caused by an intentional separation of the BI-MAS in two parts for generating the state-space report. *Dead* and *live* define two properties of CPN to check the connection between entire nodes of graph.

Table 12. State space report of BI-MAS sub-models.

BI-MAS sub-models	State space	Scc graph	Dead Transition Instances	Live Transition Instances
Single key searching - I	nodes: 37 arcs: 53 sec: 0 status: full	nodes: 37 arcs: 53 sec: 0	None	None
Single key searching-II	nodes: 562 arcs: 1024 sec: 0 status: full	nodes: 562 arcs: 1024 sec: 0	search_data_warehouse 1	None
Multi-key searching- I	nodes: 3905 arcs: 6272 sec: 2 status: full	nodes: 3905 arcs: 6272 sec: 1	None	None
Multi-key searching- II	nodes: 343 arcs: 548 sec: 0 status: full	nodes: 343 arcs: 548 sec: 0	search_data_warehouse 1	None

The second V&V method is model checking, for which Table 13 shows results. To apply this method, we use several checking properties such as reachability, detection of loops, performance peaks during run time, full system utilization, and consistent termination. These results are generated automatically based on two test cases (i.e., either single-, or multi-set scenarios of Table 11) are used as input data for the BI-MAS life-cycle.

Table 13. Model checking results for BI-MAS lifecycle.

	Modules	Loops	Performance Peaks	Utilization	Home marking	Dead marking
			Model Property			
	BI-MAS	No	compare keywords	yes	no	no
	Activate Agents	No	activating agents	yes	no	no
	Search for data-sites	No	transfer data	yes	no	no
Data-sites	Data-site1	No	search data-site 1	yes	no	no
	Data-site2	No	search data-site 2	yes	no	no
	Data-site3	No	search data-site 3	yes	no	no

For model checking, we have three separate data-sites with different names and content as listed in Table 13, to test the parallelisms in the BI-MAS. The model-checking outcome is outlined in the second column that shows no loop exists in the entire life-cycle of the BI-MAS. We prevent the loops by implementing parallel methods as depicted in Fig. 11 where three *Dispatcher* agents re activated simultaneously to search for different data-sites.

Performance peaks in Table 13 represent places in the system that are bottlenecks. Each peak requires computing power and time during execution time. Peaks exist in all sub-modules of the BI-MAS with disparate levels but their potentials are very low. For the first module of BI-MAS, a peak occurs for *compare keywords* and *ids* whether there are multi-stakeholders waiting for multi-key results. In the second sub-module *activating agents*, a peak arises while the *Facilitator* agent generates several *Dispatcher* agents based on a number of data-sites for each keyword received. For the third sub-module search for data-sites, a peak is visible during *transfer data* that is associated with a new token (time) based on a particular time sequence when the result arrives. For all three data-sites, peaks occur due to searching and comparing processes for matching several keywords within divers data sets.

The home marking that represents an initial making [34], is considered to find all initial reachable nodes relevant to the BI-MAS life-cycle. Referring to [30], the home marking can be reached from any marking state. As outlined in Table 13, no home marking exists in the defined nodes of the BI-MAS The results for checking dead markings is similar to the *Dead Transition Instances* demonstrated in Table 12. Finally, the *Utilization* test represents all the subsets of the BI-MAS are used. It means that all modules are used during the execution processes of the BI-MAS lifecycle.

6.3 Related Work Discussion

Our agent-based solution for the comprehensively designed next-generation of BI-systems is not comparable with other proposed agent-based architectures, or frameworks from literature due to the extension as stressed in the following steps.

(a) Defining a proper agent-based life-cycle for a BI-MAS with applicable functional- and non-functional requirements that are known as essential for the development of any types of agent-based BI-systems.

(b) Selection and implementation of proper agent-oriented methodologies for the development procedure of BI-MAS models.

(c) Performance of step by step analysis- and design-phases with very detailed and overall concepts for each model of a BI-MAS.

(d) Fulfillment of V&V processes including formal mapping, modeling, transformation and with the analysis results, employing different accepted checking methods.

By referring to related work of agent-based BI-systems, we reflect several consideration points that are outlined in Table 14. The first column of this table presents the list of proposed agent-based BI-systems in literature. Furthermore, in the scope of agent-oriented methodologies, a gap exists for developed agent-based BI-systems. As represented in the second column, all of these proposed BI-solutions have developed without the implementation of any specific agent-oriented methodology. The systematic analysis- and design-phases listed in the third column are equally not given for all methods. Only two of the methods are covered, while the overall phases are not define on a sufficiently detailed-level. The fourth column represents the V&V processes that are applicable only for two proposed BI-systems. As demonstrated, one of these solutions covers only simulation processes, while others cover partially experimental results related to the developed system. Overall, neither authors demonstrate the proper transformation from the analysis and design phases to implementation, nor do the authors present proper V&V results with standard tools, or methods.

Table 14. Comparison results of BI-MAS with related agent-based BI-systems

Types of agent-based BI-system	Agent-oriented methodology	Analysis & design	V&V	References
BI fusion of agent network	No	Partially covered	No	[8]
Combination Framework of BI solution Multi-agent platform (CFBM)	No	No	No	[35]
Multi Agent Based Business Intelligence (MABBI)	No	No	Simulation	[36]
Agent-based architecture of BI system	No	No	No	[37]
MAS for managing supply chains	No	No	No	[6]
Stock Trading Multi-Agent System (STMAS)	No	Partially covered	Partially covered	[11]
Self-Organized Multi-agent Technology based BI Framework	No	No	No	[38]
Model for using Agent Based Systems (ABS) in BI	No	No	No	[12]

7   Conclusion

As we represent in the content of this paper, an agent-based BI-system requires a comprehensive road-map and highly systematic development approach along with an extensive verification- and validation processes. We discover that a complex system development life-cycle comprises important phases such as analysis and design that are essential for developers. Beside the implementation of a BI-MAS life-cycle, the analysis- and design phases must ensure developers that adopted concepts such as MAS and DDM are sufficiently. To achieve this enhancement, we define the BI-MAS life-cycle along with functional and non-functional requirements for several agent-levels. To transfer each of these requirements on a system-level within the structure of MAS technology, we emphasize the need to find a comprehensive methodology. Finding an applicable agent-oriented methodology is a considerable challenge, according to literature. To tackle this challenge, we present a novel approach to evaluate nine well-known agent-oriented methodologies that shows all methodologies are incomplete.

Moreover, we also demonstrate the conceptualization processes for BI-MAS models using agent-oriented methodologies such as ROADMAP and RAP/AOR along with AOM techniques to generate a holistic development methodology. During the enactment of ROADMAP and RAP/AOR, we discover are complementary for constructing the agent-level models of a BI-MAS in the analysis- and design-phases. Additionally, several challenges and limitation occur too during these two phases. For instance, a subset of BI-MAS agents must act in a static environment and other agents are part of a dynamically changing-, or distributed environments. Representing such a distinction in models is also a challenge for other existing agent-oriented methodologies. Moreover, none-functional requirements cannot be addressed with these two methodologies, e.g., agent security is a key component in distributed environments. Adding the security concept only in a model as a quality goal in goal-model without projection into the domain model, knowledge model, etc., is not sufficient in a BI-system development life-cycle. During the implementation of a BI-MAS, a need arises for integrating complementary models to represent the processes of authenticating agents, threat detection, and so on. It is necessary to use modified methods and tools to achieve such diverse model integration.

On the other hand, our studies also discover that the CPN-tool is a good candidate with its mathematical properties to perform V&V processes of agent-based BI-systems. As V&V is the core part of development processes, we assume to perform these processes with three different methods. Besides the intended processes for V&V, CPN shows limitations while mapping and transforming BI-MAS models, e.g., the inner action of BI-MAS agents cannot be modeled using CPN-tool. On the other hand, the usage of state-space methods of CPN is generates analytical statistics about state spaces, boundedness-, home- and live-markings, and fairness properties. Consequently, a diagnostic understanding about dependability and concurrency conflicts emerges for a BI-MAS.

The current BI-MAS high-level components require exclusive tools and platforms during the implementation-phase in real life organizations. In future work, to obtain the full view of BI-MAS in enterprise-level concepts, we consider more research and extensive models, i.e. data-warehouse architecture, types of servers, types of required services, etc. In the enterprise-level deployments of BI-MAS again, there exists a need for further research including two parts. Firstly, for the definition of the entire contextual organizational structure on a system-level, the integration of data warehouses and analytics tools requires additional research work. Secondly, describing and developing user-interfaces, middleware applications, and secure protocols are the second part that needs research and development work. Unclear is also the projection of important non-functional requirements such as security into other model types, e.g., for agent behavior and –interaction. Finally, the verification- and simulation capabilities of CPN do not cover all aspects of a BI-MAS to address dependability issues and concurrency conflicts. Thus, we plan to explore additional formal checking techniques for the goal of highly relyable system development.

References

1. Bologa, A.-R., Bologa, R.: Business intelligence using software agents. Database Syst. J. 2 (4), 31–42 (2011)
2. Draheim, D.: Smart business process management. In: 2011 BPM and Workflow Handbook, Digital Edition. Future Strategies, Workflow Management Coalition, pp. 207–223 (2012)
3. Atkinson, C., Draheim, D.: Cloud-aided software engineering: evolving viable software systems through a web of views. In: Mahmood, Z., Saeed, S. (eds.) Software Engineering Frameworks for the Cloud Computing Paradigm. Springer, London (2013). https://doi.org/10.1007/978-1-4471-5031-2_12
4. Mabrouk, T.F., El-Sherbiny, M.M., Guirguis, S.K., Shawky, A.Y.: A multi-agent role-based system for business intelligence. In: Sobh, T. (ed.) Innovations and Advances in Computer Sciences and Engineering. Springer, Dordrecht (2010). https://doi.org/10.1007/978-90-481-3658-2_35
5. Carbonell, J.G., Siekmann, J.: Multi-Agent Systems and Applications III. Springer, Heidelberg (2005). https://doi.org/10.1007/3-540-45023-8
6. Hemamalini, R., Mary, L.J.: An analysis on multi - agent based distributed data mining system. Int. J. Sci. Res. Publ. 4(6), 1–6 (2014)
7. Devi, S.: A survey on distributed data mining and its trends. Int. J. Res. Eng. Technol. 2(3), 107–120 (2014)
8. Zeng, L., et al.: Distributed data mining: a survey. Inf. Technol. Manage. 13(4), 403–409 (2012)
9. Salih, N.K., Zang, T., Viju, G.K., Mohamed, A.A.: Autonomic management for multi-agent systems. IJCSI Int. J. Comput. Sci. Issues 8(5), 338–341 (2011)
10. Lee, C., Lau, H., Ho, G., Ho, W.: Design and development of agent-based procurement system to enhance business intelligence Expert syst. Appl. 36, 877–884 (2009)
11. Khozium, M.O.: Multi-agent system overview: architectural designing using practical approach. Int. J. Comput. Technol. 5(2), 85–93 (2013)
12. Kargupta, H., Hamzaoglu, I., Stafford, B.: Scalable, distributed data mining-an agent architecture. In: Proceedings Third International Conference on Knowledge Discovery and Data Mining, pp. 211–214 (1997)

13. Krishnaswamy, S., Zaslavsky, A., Loke, S.W.: An architecture to support distributed data mining services in e-commerce environments. In: Advanced Issues of E-Commerce and Web-Based Information Systems, WECWIS 2000, pp. 239–246 (2000)
14. Sen, S.K., Dash, S., Pattanayak, S.P.: Agent based meta learning in distributed data mining system. Int. J. Eng. Res. Appl. (IJERA) 2(3), 342–348 (2012)
15. Loebbert, A., Finnie, G.: A multi-agent framework for distributed business intelligence systems. In: 45th Hawaii International Conference on System Sciences (2012)
16. Ghandehari, E., Saadatjoo, F., Chahooki, M.A.Z.: Method integration: an approach to develop agent oriented methodologies. J. Artif. Intell. Data Min. 3(1), 59–76 (2015)
17. Sterling, L.S., Taveter, K.: The Art of Agent-Oriented Modeling. MIT Press Ebooks, Cambridge (2009)
18. Bresciani, P., Perini, A., Giogini, P., Giunchiglia, F., Mylopoulos, J.: Tropos: an agent-oriented software development methodology. Auton. Agent. Multi-Agent Syst. 8, 203–236 (2004)
19. Cossentino, M., Gaglio, S., Sabatucci, L., Seidita, V.: The PASSI and agile PASSI MAS meta-models compared with a unifying proposal. In: Pěchouček, M., Petta, P., Varga, L.Z. (eds.) CEEMAS 2005. LNCS (LNAI), vol. 3690, pp. 183–192. Springer, Heidelberg (2005). https://doi.org/10.1007/11559221_19
20. Juan, T., Pearce, A., Sterling, L.: ROADMAP: extending the Gaia methodology for complex open system. In: The first International Joint Conference on Autonomous Agents and Multiagent System: Part 1, pp. 3–10. ACM (2002)
21. Desouza, K.C.: Intelligent agents for competitive intelligence: survey of applications. Compet. Intell. Rev. 12(4), 57–63 (2001)
22. Chaudhuri, S., Dayal, U., Narasayya, V.: An overview of business intelligence technology. Commun. ACM 54(8), 88–98 (2011)
23. Expert OLAP.com. http://olap.com/learn-bi-olap/olap-bi-definitions/business-intelligence/. Accessed 21 Dec 2017
24. Ozlszak, C.M., Ziemba, E.: Approach to building and implementing business intelligence system. Interdiscip. Jo. Inf. Knowl. Manag. 2, 135–148 (2007)
25. Ranjan, J.: Business intelligence: concept, components, techniques and benefits. J. Theor. Appl. Inf. Technol. 9(1), 60–70 (2009)
26. Matillion: What businesses really want from business intelligence and Analytics (2017). www.matillion.com
27. Liu, S.: Business Intelligence Fusion Based on Multi-agent and Complex Network. J. Softw. 9(11), 2804–2812 (2014)
28. Bobek, S., Perko, I.: Intelligent agent based business intelligence. In: Current Developments in Technology-Assisted Education, pp. 1047–1051 (2006)
29. Hevner, A.R., March, S.T., Park, J., Ram, S.: Design science in information systems research. MIS Q. 28(1), 75–105 (2014)
30. Dam, K.H., Winikoff, M.: Comparing agent-oriented methodologies. In: Giorgini, P., Henderson-Sellers, B., Winikoff, M. (eds.) AOIS -2003. LNCS (LNAI), vol. 3030, pp. 78–93. Springer, Heidelberg (2004). https://doi.org/10.1007/978-3-540-25943-5_6
31. A. Sturm and O. Shehory, "A framework for evaluating agent-oriented methodologies," Agent-Oriented Information Systems, Springer, pp. 94–109, 2004
32. Sturm, A., Shehory, O.: A framework for evaluating agent-oriented methodologies. In: Giorgini, P., Henderson-Sellers, B., Winikoff, M. (eds.) AOIS -2003. LNCS (LNAI), vol. 3030, pp. 94–109. Springer, Heidelberg (2004). https://doi.org/10.1007/978-3-540-25943-5_7

33. Herlea, D.E., Jonker, C.M., Treur, J., Wijngaards, N.J.E.: Specification of bahavioural requirements within compositional multi-agent system design. In: Garijo, Francisco J., Boman, M. (eds.) MAAMAW 1999. LNCS (LNAI), vol. 1647, pp. 8–27. Springer, Heidelberg (1999). https://doi.org/10.1007/3-540-48437-X_2

34. Padgham, L., Winikoff, M.: Prometheus: a methodology for developing intelligent agents. In: Agent-Oriented Software Engineering III, pp. 174–185 (2003)

35. Bernon, C., Gleizes, M.-P., Picard, G., Glize, P.: The ADELFE methodology for an intranet system design. In: Equipe SMAC; Systèmes Multi-Agents Coopératifs (2002)

36. Tran, Q.-N.N., Low, G.: MOBMAS: a methodology for ontology-based multi-agent systems development. Inf. Softw. Technol. 50(7–8), 697–722 (2008)

37. Soleimanian, F., Zabardast, B., Amini, E.: Analysis and design by agent based MaSE methodology: a case study. Int. J. Comput. Appl. 63(4), 10–15 (2013)

38. Wooldridge, M., Jennings, N.R., Kinny, D.: The Gaia mthodology for agent-oriented analysis and design. In: JAAMAS, pp. 1–27 (2000)

39. Zamboell, F., Jennings, N.R., Wooldridge, M.: Developing multiagent systems: the Gaia methodology. ACM Trans. Softw. Eng. Methodol. (TOSEM) 12(3), 317–370 (2003)

40. Taveter, K., Wagner, G.: Agent-oriented modeling and simulation of distributed manufacturing. Idea Group Inc., pp. 1–14 (2007)

41. Domann, J., Hartmann, S., Burkhardt, M., Barge, A., Albayrak, S.: An agile method for multiagent software engineering. In: The 1st International Workshop on Developing and Applying Agent Framework (DAAF), pp. 928–934. ELSEVIER (2014)

42. Norta, A., Yangarber, R., Carlson, L.: Utility evaluation of tools for collaborative development and maintenance of ontologies. In: 2010 14th IEEE International on Enterprise Distributed Object Computing Conference Workshops (EDOCW), pp. 207–214 (2010)

43. Taveter, K.: Towards radical agent-oriented software engineering processes based on AOR modelling. In: Agent-oriented methodologies, Idea Group Inc., pp. 277–316 (2005)

44. Jennings, N.R., Norman, T.J., Faratin, P., O'Brien, P., Odgers, B.: Autonomous agents for business process management. Appl. Artif. Intell. 14(2), 145–189 (2000)

45. Al-Neaimi, A., Qatawneh, S., Saiyd, N.A.: Conducting verification and validation of multi-agent systems. arXiv preprint arXiv:1210.3640 (2012)

46. Cheng, Betty H.C., et al.: Software engineering for self-adaptive systems: a research roadmap. In: Cheng, B.H.C., de Lemos, R., Giese, H., Inverardi, P., Magee, J. (eds.) Software Engineering for Self-Adaptive Systems. LNCS, vol. 5525, pp. 1–26. Springer, Heidelberg (2009). https://doi.org/10.1007/978-3-642-02161-9_1

47. Jensen, K., Kristensen, L.M.: Coloured Petri Nets: Modelling and Validation of Concurrent Systems. Springer, Heidelderg (2009). https://doi.org/10.1007/b95112

48. Honby, G.S.: Measuring, enabling and comparing modularity, regularity and hierarchy in evolutionary design. In: Proceedings of the 7th Annual Conference on Genetic and Evolutionary Computation, pp. 1729–1736. ACM (2005)

49. Donatelli, S.: Petri Nets and Other Models of Concurrency. Springer, Turku (2006). https://doi.org/10.1007/978-3-662-55862-1

50. Venable, J., Pries-Heje, J., Baskerville, R.: FEDS: a framework for evaluation in design science research. Eur. J. Inf. Syst. 25, 77–89 (2016)

51. Norta, A.: Creation of smart-contracting collaborations for decentralized autonomous organizations. In: Matulevičius, R., Dumas, M. (eds.) BIR 2015. LNBIP, vol. 229, pp. 3–17. Springer, Cham (2015). https://doi.org/10.1007/978-3-319-21915-8_1

52. Kutvonen, L., Norta, A., Ruohomaa, S.: Inter-enterprise business transaction management in open service ecosystems. In: 2012 IEEE 16th International Enterprise Distributed Object Computing Conference (EDOC), pp. 31–40 (2012)
53. Thai, T.M., Amblard, F., Gaudou, B.: Combination framework of BI solution \& multi-agent platform (CFBM) for multi-agent based simulations. In: 3EME Conference francophone sur le Gestion et l'Extraction de Connaissances: Journée Atelier aide à la Décision à tous les, Etages (AIDE@ EGC 2013), pp. 35–42 (2013)
54. Sperka, R.: Agent-based design of business intelligence system architecture. J. Appl. Econ. Sci. **VII**((3(21)) (2012)
55. Venkatadri, M., Sastry, M.G., Manjunath, G.: A novel business intelligence system framework. Univers. J. Comput. Sci. Eng. Technol. **1**, 112–116 (2010)
56. Rao, V.S.: Multi agent-based distributed data mining : an overview. Int. J. Rev. Comput. (2076–3328), 82–92 (2010)

Enhancing Rating Prediction Quality Through Improving the Accuracy of Detection of Shifts in Rating Practices

Dionisis Margaris[1] and Costas Vassilakis[2(✉)]

[1] Department of Informatics and Telecommunications, University of Athens,
Athens, Greece
margaris@di.uoa.gr
[2] Department of Informatics and Telecommunications,
University of the Peloponnese, Tripoli, Greece
costas@uop.gr

Abstract. The most widely used similarity metrics in collaborative filtering, namely the Pearson Correlation and the Adjusted Cosine Similarity, adjust each individual rating by the mean of the ratings entered by the specific user, when computing similarities, due to the fact that users follow different rating practices, in the sense that some are stricter when rating items, while others are more lenient. However, a user's rating practices change over time, i.e. a user could start as lenient and subsequently become stricter or vice versa; hence by relying on a single mean value per user, we fail to follow such shifts in users' rating practices, leading to decreased rating prediction accuracy. In this work, we present a novel algorithm for calculating dynamic user averages, i.e. time-in-point averages that follow shifts in users' rating practices, and exploit them in both user-user and item-item collaborative filtering implementations. The proposed algorithm has been found to introduce significant gains in rating prediction accuracy, and outperforms other dynamic average computation approaches that are presented in the literature.

Keywords: Recommender systems · Collaborative filtering
User-user similarity · Item-item similarity · Dynamic average
Prediction accuracy · Ratings' timestamps

1 Introduction

Collaborative filtering (CF) computes personalized recommendations, by taking into account users' past likings and tastes, in the form of ratings entered in the CF rating database. User-user CF algorithms firstly identify people having similar tastes, by examining the resemblance of already entered ratings; for each user u, other users having highly similar tastes with u are designated as u's nearest neighbors (NNs). Afterwards, in order to predict the rating that u would give to an item i that u has not reviewed yet, the ratings assigned to item i by u's NNs are combined [1], under the assumption that users are highly likely to exhibit similar tastes in the future, if they have done so in the past as well [26, 30]. Analogous practices are followed in item-item

© Springer-Verlag GmbH Germany, part of Springer Nature 2018
A. Hameurlain and R. Wagner (Eds.): TLDKS XXXVII, LNCS 10940, pp. 151–191, 2018.
https://doi.org/10.1007/978-3-662-57932-9_5

CF algorithms, where the first step is to locate items that are similarly rated by users. CF is the most successful and most applied technique in the design of recommender systems [3]. In order to measure similarity between users or items, the Pearson Correlation Coefficient and the Adjusted Cosine Similarity [4] (see also Sect. 3) are the most commonly used formulas in CF recommender systems. In this context, both the Pearson correlation coefficient and the Adjusted Cosine Similarity adjust the ratings of a user u by the mean value of all ratings entered by u, and the ratings of an item i by the mean value of all ratings for this item, respectively, so as to tackle the issue that some users may rate items higher than others or that some items may be rated higher than others. However, relying on a single, global mean value presumes that the users' marking practices remain constant over time; in practice though, it is possible that a user's marking practices change over time, i.e. a user could start off being strict and subsequently change to being lenient, or vice versa; similarly, an item could start as being high rated and subsequently begin receiving lower marks, due to general shift of interest (music or clothing trends, decline of interest for blockbuster movies or books etc.). For instance, according to the MovieLens 20M dataset [9, 10], the Titanic movie started off with an average of 4.30/5 in 1997, dropping to 3.06 in 2005, and finally climbing back to 3.24 in 2015.

Similar situations arise for users: for example, consider that a user initially grades *Tudors* (http://www.imdb.com/title/tt0758790/), which is the first historic period drama series of high quality that she rates; being enthusiastic with the series, she enters a rating of 10. Subsequently, the same user rates *Game of Thrones* (http://www.imdb.com/title/tt0944947/), which she finds superb and better than *Tudors*, giving it the highest available mark, i.e. 10. Finally, the user watches a few episodes from the show *Vikings* (http://www.imdb.com/title/tt2306299), and grades this series with an 8. While both *Tudors* and *Game of Thrones* have been equally rated by the user, this does not necessarily reflect the fact that she considers them of equal quality; similarly, the fact that *Vikings* got a lower grade than *Tudors*, does not necessarily mean that she considers it of inferior quality: the user's rating criteria and practices have simply evolved, along with her experiences on historic period drama series.

Insofar, while many efforts have been made to improve the CF prediction accuracy, and the aspect of changes in users' interests has been extensively studied (Gama et al. [24] provide a comprehensive review), the issue of shifts in rating practices has not received adequate attention. Margaris and Vassilakis [33] introduce the concept of dynamic user rating averages which follow the users' marking practices shifts and present two alternative algorithms for computing a user's dynamic averages. These algorithms are validated in the context of user-user CF, and have been found to achieve better rating prediction accuracy than the plain CF algorithm, using the Pearson correlation similarity metric.

In this paper, we extend the work in [33] by introducing a new dynamic average computation algorithm, namely DA_{next}, which is capable of better following the users' marking practices shifts, leading to improved prediction accuracy, as compared to the two dynamic average algorithms presented in [33]. This improvement is consistent under both user-user CF implementations and item-item CF implementations, where similarities are measured using the Pearson Correlation and the Adjusted Cosine Similarity respectively. To validate our approach, we present an extensive comparative evaluation among (i) the proposed algorithm, (ii) the two approaches proposed in [33] and (iii) the classic static average (unique mean value), considering both the user-user and the item-item CF implementations.

The proposed algorithm is based on the exploitation of timestamp information which is associated with ratings; hence in this work, we use the Amazon datasets [7, 8], the MovieLens datasets [9, 10], the Netflix dataset [11] and the Ciao dataset [51], which contain timestamps. It is worth noting that the proposed algorithm can be combined with other techniques that have been proposed for either improving prediction accuracy in CF-based systems, including consideration of social network data (e.g. [14, 25, 29]), location data (e.g. [34, 35]) and pruning of old user ratings (e.g. [12, 38]), or techniques for speeding up prediction computation time, such as clustering (e.g. [36, 37, 40]).

The rest of the paper is structured as follows: Sect. 2 overviews related work, while Sect. 3 presents the proposed algorithm, together with the algorithms presented in [33], for self-containment purposes. Section 4 evaluates the proposed algorithm using the aforementioned datasets and finally, Sect. 5 concludes the paper and outlines future work.

2 Related Work

The accuracy of CF-based systems is a topic that has attracted considerable research efforts. Koren [15] proposes a new neighborhood-based model, which is based on formally optimizing a global cost function and leads to improved prediction accuracy, while maintaining merits of the neighborhood approach such as explainability of predictions and ability to handle new ratings (or new users) without retraining the model. In addition, he suggests a factorized version of the neighborhood model, which improves its computational complexity while retaining prediction accuracy. Liu et al. [18] present a new user similarity model to improve the recommendation performance when only few ratings are available to calculate the similarities for each user. The model considers the local context information of user ratings, as well as the global preference of user behavior. Ramezani et al. [39] propose a method to find the neighbor users based on the users' interest patterns in order to overcome challenges like sparsity and computational issues, following the idea that users who are interested in the same set of items share similar interest patterns, therefore, the non-redundant item subspaces are extracted to indicate the different patterns of interest and then, a user's tree structure is created based on the patterns he has in common with the active user.

Research has shown that exploiting time in the rating prediction computation can improve prediction accuracy, due to concept drift; concept drift is the phenomenon when the relation between the input data and the target variable changes over time [24]. Change of interests [5, 24] is a typical example of concept drift. To this end, Zliobaite et al. [22] develop an intelligent approach for sales prediction, which uses a mechanism for model switching, depending on the sales behavior of a product. This research presents an intelligent two level sales prediction approach that switches the predictors depending on the properties of the historical sales. This approach is shown to achieve better results as compared to both a baseline predictor and an ensemble of predictors.

Ang et al. [21] address the problem of adaptation when external changes are asynchronous, by developing an ensemble approach, called PINE, which combines reactive adaptation via drift detection, and proactive handling of upcoming changes via early warning and adaptation across the peers. In addition, PINE is parameter-insensitive and incurs less communication cost while achieving better accuracy.

Elwell and Polikar [20] tackle the issue of concept drift in the context of online learning, introducing a batch-based ensemble of classifiers, called Learn++.NSE, where NSE stands for Non-Stationary Environments. Learn++.NSE learns from consecutive batches of data without making any assumptions on the nature or rate of drift; it can learn from such environments that experience constant or variable rate of drift, addition or deletion of concept classes, as well as cyclical drift. The algorithm learns incrementally, as do other members of the Learn++ family of algorithms, that is, without requiring access to previously seen data. Learn++.NSE trains one new classifier for each batch of data it receives, and combines these classifiers using a dynamically weighted majority voting. The algorithm is evaluated on several synthetic datasets designed to simulate a variety of nonstationary environments, as well as a real-world weather prediction dataset. Minku et al. [19] present a new categorization for concept drift, separating drifts according to different criteria into mutually exclusive and non-heterogeneous categories. Moreover, they present a diversity analysis in the presence of different types of drifts and it shows that, before the drift, ensembles with less diversity obtain lower test errors. Nishida and Yamauchi [17] have developed a detection method that includes an online classifier and monitors its prediction errors during the learning process, which uses a statistical test of equal proportions. Experimental results showed that this method performed well in detecting the concept drift in five synthetic datasets that contained various types of concept drift.

Vaz et al. [16] propose an adaptation of the item-based CF algorithm to incorporate rating age influence in predictions. It considers ratings in two dimensions: the active user ratings and the community ratings, and it inserts a time weight, which gives more relevance to more recent ratings than to older ones, both in the similarity calculation and in the rating prediction equation.

Dror et al. [2] consider the temporal dimension in the context of recommender systems by capturing different temporal dynamics of music ratings, along with information from the taxonomy of music-related items; both these dimensions are exploited by a rich bias model. The method proposed in this work is applied on a sparse, large-scale dataset, and the particular characteristics of the dataset are extracted and utilized. Liu et al. [13] present a social temporal collaborative ranking model that can simultaneously achieve three objectives: (1) the combination of both explicit and implicit user feedback, (2) support for time awareness using an expressive sequential matrix factorization model and a temporal smoothness regularization function to tackle overfitting, and (3) support for social network awareness by incorporating a network regularization term. Dias and Fonseca [31] explore the usage of temporal context and session diversity in session-based CF techniques for music recommendation. They compare two techniques to capture the users' listening patterns over time: one explicitly extracts temporal properties and session diversity, to group and compare the similarity of sessions, the other uses a generative topic modeling algorithm, which is able to implicitly model temporal patterns. Results reveal that the inclusion of temporal information, either explicitly or implicitly, significantly increases the accuracy of the recommendation, as compared to the traditional session-based CF.

Li et al. [41] study the problem of predicting the popularity of social multimedia content embedded in short microblog messages, exploiting the idea of concept drift to capture the phenomenon that through the social networks' "re-share" feature, the popularity of a multimedia item may be revived or evolve. They model the social multimedia item popularity prediction problem using a classification-based approach which is used for two sub-tasks, namely re-share classification and popularity score classification. Furthermore, they develop a concept drift-based popularity predictor by ensembling multiple trained classifiers from social multimedia instances in different time intervals.

Lu et al. [42] present a novel evolutionary view of user's profile by proposing a Collaborative Evolution (CE) model, which learns the evolution of user's profiles through the sparse historical data in recommender systems and outputs the prospective user profile of the future. Kangasrääsiö et al. [43] formulate a Bayesian regression model for predicting the accuracy of each individual user feedback and thus find outliers in the feedback data set. Additionally, they introduce a timeline interface that visualizes the feedback history to the user and provides her with suggestions on which past feedback is likely in need of adjustment. This interface also allows the user to adjust the feedback accuracy inferences made by the model. The proposed modeling technique, combined with the timeline interface, makes it easier for the users to notice and correct mistakes in their feedback, and to discover new items.

Lo et al. [52] address the issue of tracking concept drift in individual user preferences; in this context they develop a Temporal Matrix Factorization approach (TMF) for tracking concept drift in each individual user latent vector. To this end, a time series of rating matrices is initially constructed from the ratings database; subsequently this time series is used to capture the concept drift dynamics for each individual user; and finally, the captured dynamics are taken into account in the rating prediction computation phase. Cheng et al. [53] propose the ISCF (interest sequences CF), a recommendation method based on users' interest sequences; interest sequences are first detected from the ratings, and are subsequently used to refine similarity metrics between users, thus taking into account dynamic evolution patterns of users' preferences.

However, none of the above mentioned works considers the issue of shifts in the users' rating practices. This issue has only recently received some attention: Margaris and Vassilakis [33] introduce and exploit the concept of dynamic user rating averages which follow the users' marking practices shifts. Furthermore, they present two alternative algorithms for computing a user's dynamic averages and perform a comparative evaluation in the context of a user-user CF implementation. The results of this evaluation show that the dynamic average-based algorithms exhibit better performance than the plain CF algorithm in terms of rating prediction accuracy, at the expense of a small to tolerable drop in coverage.

Interestingly, fuzzy recommender systems (FRS) [50] introduce the concept of fuzzy user context in the process of rating prediction and recommendation formulation. Under this approach, each rating entered by a user is associated with a particular context element through a fuzzy membership function. The FRS approach could be exploited for accommodating user rating practices as a specific rating criterion; subsequently the criterion would be used in the calculation of fuzzy similarity degree between users and finally be incorporated in the final rating prediction. To implement this approach would necessitate however a concrete, automated method for assigning strictness labels to individual user ratings, the definition of an appropriate membership & utility functions, and the evaluation of the overall system performance. To our knowledge, no FRS has been reported in the literature to accommodate these features.

This paper extends the work presented in [33] by (1) introducing a novel algorithm for dynamic average computation, which is able to follow the users' shifts in rating practices more accurately and (2) validating its performance against widely used datasets with diverse characteristics, exploring both the user-user and item-item CF implementations. The DA_{next} algorithm introduced in this paper is based on the rationale that the rating practices of a user at some time point is formulated according to the experiences she has amassed up to that time point and therefore it is more accurately reflected by her subsequent ratings, which are influenced by the same (and some additional) experiences. On the other hand, the $DA_{vicinity}$ algorithm introduced in [33] assumes that the user's ratings are mostly affected by temporally constrained factors, such as user mood, while the $DA_{previous}$ algorithm, also presented in [33], assumes that the rating-related behavior of a user at a certain time point is better estimated by considering the user's behavior up to that time point. While it is also possible that the ratings entered by a user during some period are affected by her mood [47], which would favor the $DA_{vicinity}$ algorithm, the results indicate that the effect of the amassed experiences is stronger than the effect of mood. Further qualitative evaluation on this subject is required, and this is envisioned as part of our future work.

The newly introduced algorithm has been found to provide more accurate rating predictions, by better capturing the shifts in users' rating practices.

3 Exploiting Ratings' Timestamps in Users Dynamic Average Configuration

In CF, predictions for a user X are computed based on a set of users which have rated items similarly with X; this set of users is termed "near neighbors of X" (X's NNs). The predominant similarity metric used in CF-based systems is the Pearson correlation metric [3], where the similarity between two users X and Y is expressed as:

$$Pearson\_sim(X, Y) = \frac{\sum_{i \in I_X \cap I_Y} (R_{X,i} - \overline{R_X}) * (R_{Y,i} - \overline{R_Y})}{\sqrt{\sum_{i \in I_X \cap I_Y} (R_{X,i} - \overline{R_X})^2} * \sqrt{\sum_{i \in I_X \cap I_Y} (R_{Y,i} - \overline{R_Y})^2}} \quad (1)$$

where i ranges over items that have been rated by both X and Y and $\overline{R_X}$ (resp. $\overline{R_Y}$) is the mean value of the ratings entered by X (resp. Y); as noted above, the Pearson correlation formula uses a "global" mean value. The algorithms presented in this section target the computation of $\overline{R_X}$ (resp. $\overline{R_Y}$), aiming to substitute the global average, which is insensitive to shifts in rating practices, by an average that is tailored to the time period that $R_{X,i}$ (resp. $R_{Y,i}$) was entered. When a dynamic average computation algorithm DA_{Alg} is employed, the above formula is modified as:

$$Pearson\_sim(X,Y) = \frac{\sum_{i \in I_X \cap I_Y}(R_{X,i} - DA_{Alg}(R_{X,i})) * (R_{Y,i} - DA_{Alg}(R_{Y,i}))}{\sqrt{\sum_{i \in I_X \cap I_Y}(R_{X,i} - DA_{Alg}(R_{X,i}))^2} * \sqrt{\sum_{i \in I_X \cap I_Y}(R_{Y,i} - DA_{Alg}(R_{Y,i}))^2}}$$

(2)

Similarly, for item-item CF, the Adjusted Cosine Similarity (which is preferred against the basic cosine similarity metric, since it takes into account the differences in rating scale between different users [46]) is modified as:

$$Adj\_cos\_sim(i,j) = \frac{\sum_{u \in U}(R_{u,i} - DA_{Alg}(R_{u,i})) * (R_{u,j} - DA_{Alg}(R_{u,j}))}{\sqrt{\sum_{u \in U}(R_{u,i} - DA_{Alg}(R_{u,i}))^2} * \sqrt{\sum_{u \in U}(R_{u,j} - DA_{Alg}(R_{u,j}))^2}}$$

(3)

where u ranges over the users that have rated both i and j; again $DA_{Alg}(R_{u,i})$ denotes the dynamic average of user u at the time that $DA_{Alg}(R_{u,i})$ was submitted.

While other similarity metrics, such as Euclidian distance [49], Manhattan Distance [49], Spearman's coefficient [4], Kendall's Tau [4] etc. have been used in recommender systems, in this paper we will confine ourselves to examining the Pearson similarity metric for the user-user strategy and the adjusted cosine similarity metric for the item-item strategy. This is due to the fact that the relevant formulas readily use the average values of the users' and items' ratings, hence the substitution of the global user's or item's average by the rating-specific dynamic average is a natural extension, while other similarity metrics do not adjust ratings according to the global average. This is also the case with the very promising matrix factorization technique [32]. In our future work, we plan to investigate how dynamic averages can be integrated into the above mentioned methods.

In the rest of this section we present the proposed technique for computing the dynamic user averages. For completeness purposes, we will also describe the relevant techniques described in [33], which are also used as yardsticks in the performance evaluation section.

3.1   The Proposed Algorithm

Under the proposed approach for computing dynamic averages, a separate average for each rating is calculated and stored. The algorithm for computing the dynamic average proposed in this paper takes into account only the ratings that have been submitted *after* the rating for which the dynamic average is submitted. Effectively, this algorithm is based on the assumption that, when considering a particular rating r, ratings that have been entered after r reflect more accurately the user's strictness at the time point that r was entered. Under this approach for computing dynamic averages, each user rating $r_{u,x}$ is coupled with its own dynamic average $DA_{next}(r_{u,x})$ which is computed as shown in Eq. 4:

$$DA_{next}(r_{u,x}) = \frac{\sum_{r \in Ratings(u) \bigwedge t(r) > t(r_{u,x})} r}{|r : r \in Ratings(u) \bigwedge t(r) > t(r_{u,x})|} \tag{4}$$

The pseudocode for the computation of the dynamic averages under the proposed algorithm is illustrated in Listing 1.

```
PROCEDURE bootstrapDynamicAverages(ratingsDB)
// Input: ratings database containing all users' ratings
// Output: the ratings database is complemented with
//    the dynamic averages of the ratings. For each rating,
//    the number of subsequent ratings is also stored to
//    facilitate the update of dynamic averages

FOREACH user u ∈ users(ratingsDB)
    ru = retrieveAllUserRatings(u, ratingsDB)
    ru = sort ru by rating timestamp in asc order
    calculateDAnext(ru)
END FOR
END PROCEDURE


PROCEDURE calculateDAnext(ratingList)
// Input: a list of ratings in ascending temporal order.
// Output: the dynamic averages of all the ratings in the
// input list have been caclulated

// the last rating has no next, so its dynamic average
// defaults to the rating itself
numRatings = count(ratingList)
ratingList[numRatings].dynamicAVG = ratingList[numRatings].rating

sumOfNextRatings = 0
FOR i = numRatings -1 DOWNTO 1 STEP -1
    sumOfNextRatings += ratingList[i + 1].rating
    ratingList[i].dynamicAVG = (sumOfNextRatings / (numRatings - i))
END FOR
END PROCEDURE
```

Listing 1. Pseudocode for the computation of the dynamic averages in the ratings database

After the dynamic averages have been computed as illustrated in Listing 1, the Pearson similarity between two users X and Y can be computed as shown in Listing 2, which implements the formula given in Eq. 2. The computation between each pair of users can be done while the algorithm bootstraps, and the cached similarities can be used thereafter for rating prediction, as in the typical case of user-user CF-based systems.

```
FUNCTION DA_BasedPearsonSimilarity(RatingsDB, X, Y)
// Input: ratings database containing all users' ratings and the
// identities of the users
// Output: Pearson similarity metric

ratingsX = retrieveAllUserRatings(X, RatingsDB)
ratingsY = retrieveAllUserRatings(Y, RatingsDB)
// find ids of items rated by both users
commonItemIDs = commonItems(ratingsX, ratingsY)
pearsonNominator = 0
pearsonDenomX = 0
pearsonDenomY = 0
FOREACH itemID in commonItemIDs
  ratingX = getRatingByItemID(ratingsX, itemId)
  ratingY = getRatingByItemID(ratingsX, itemId)
  pearsonNominator += (ratingX.rating - ratingX.dynamicAVG) *
                      (ratingY.rating - ratingY.dynamicAVG)
  pearsonDenomX += pow((ratingX.rating - ratingX.dynamicAVG), 2)
  pearsonDenomY += pow((ratingY.rating - ratingY.dynamicAVG), 2)
END FOR
RETURN pearsonNominator / (sqrt(pearsonDenomX) *
                           sqrt(pearsonDenomY))
END FUNCTION
```

Listing 2. Pseudocode for the computation of the Pearson similarity between two users, considering the dynamic averages

When a new rating is entered in the database by some user u, the denominator of Eq. 4 changes for all ratings r that have been entered by the particular user, therefore the dynamic averages for all ratings entered by u must be recalculated. This will in turn trigger the recalculation of all similarities between u and other users in a user-user CF implementation (cf. Eq. 2) or the recalculation of all similarities between items that u has rated and other items in an item-item CF implementation (cf. Eq. 3). The relevant performance implications are discussed in Subsect. 4.10, together with the memory and secondary storage requirements of the algorithm.

```
PROCEDURE addRating(ratingsDB, user, item, rating)
// add a new rating in the ratings database

// INPUT: rating database, the user that gave the rating,
// the item on which the rating was given and the rating
// value
// Output: updated rating database with the user's new
// dynamic averages
// append rating to ratings db; this is positioned last, main-
taining
// the temporal sort order within the ratings of each user
// that has been established in the algorithm bootstrap
appendRating(ratingsDB, new Rating(user, item, rating))
ru = retrieveAllUserRatings(user, ratingsDB)
// recalculate the user's dynamic averages
calculateDAnext(ru)
END PROCEDURE
```

Listing 3. Pseudocode for inserting a new rating into the ratings database

One issue that is worth discussing in the dynamic average computation procedure described above is that recent ratings have only few next ones, therefore the dynamic average computed for these ratings may be skewed. While this is true, it has to be noted that the most recent ratings of each user u, where this skew appears, are only a small fraction of the items that u has in common with her near neighbors (in our experiments, less than 3.6% of the computations for evaluating similarities between users involved the last four ratings of either u or u's near neighbors), with the rest of the computations being based on previous ratings that have at least four next ratings. In this respect, the effect of this skew is small. In order to further improve the effectiveness of the algorithm and minimize skew, variations of the algorithm may be introduced, which would e.g. consider a number of *past* ratings in the dynamic average computation or use the global average when an adequate number of more recent ratings is not available. Further elaboration and experimentation on this aspect is required, and this is considered part of our future work.

3.2   Existing Dynamic Average Algorithms

In [33], two algorithms for computing dynamic user averages were proposed, namely (a) the dynamic average based on the temporal vicinity of the ratings, which will be denoted as $DA_{vicinity}$ and (b) the dynamic average based only on previous ratings, which will be denoted as $DA_{previous}$. While Margaris and Vassilakis [33] describe their application only in a user-user CF scenario, these algorithms can be also directly applied in an item-item CF implementation, by using the corresponding dynamic averages in Eq. 3. In the next subsections, we briefly present these algorithms for completeness purposes.

Computing the Dynamic Average Based on the Temporal Vicinity of the Ratings. According to the $DA_{vicinity}$ algorithm, when computing the dynamic average $DA_{vicinity}(r)$ for a rating r, each user rating r' posted by the same user is assigned a weight on the basis of its temporal vicinity to r: ratings that have been entered temporally close to r are assigned higher weights, and as temporal distance increases, the weights decrease. This approach is based on the rationale that user ratings that are temporally distant to R may not accurately reflect the user's strictness at that particular time, while ratings that are temporally close to r form a better basis for deriving user strictness for the time period that r was entered.

In more detail, for rating $r_{u,x}$ of item x by user u, which has been entered at $t(r_{u,x})$, the weight $w_{u,x}(r_{u,i})$ of a rating $r_{u,i}$ is computed using the standard normalization function presented in [27]:

$$w_{u,x}(r_{u,i}) = 1 - \frac{|t(r_{u,x}) - t(r_{u,i})|}{\max_{i \in Ratings(u)}(t(r_{u,i})) - \min_{i \in Ratings(u)}(t(r_{u,i}))} \tag{5}$$

where $t(r_{u,i})$ is the timestamp of rating $r_{u,i}$, whereas $\min_{i \in Ratings(u)}(t(r_{u,i}))$ and $\max_{i \in Ratings(u)}(t(r_{u,i}))$ denote the minimum and the maximum timestamp in the database among ratings entered by user u, respectively.

Finally, the dynamic average associated to rating $r_{u,x}$ is computed using the formula:

$$DA_{vicinity}(r_{u,x}) = \frac{\sum_{r \in Ratings(u)} w_{u,x}(r) * r}{\sum_{r \in Ratings(u)} w_{u,x}(r)} \tag{6}$$

Computing the Dynamic Average Based only on Previous Ratings. Under this approach for computing dynamic averages, again each user rating $r_{u,x}$ is coupled with its own average $DA_{previous}(r_{u,x})$. When computing this average, only ratings entered by the same user (u) prior to $r_{u,x}$ are taken into account; formally this approach is expressed by Eq. 7:

$$DA_{previous}(r_{u,x}) = \frac{\sum_{r \in Ratings(u) \wedge t(r) < t(r_{u,x})} r}{|r : r \in Ratings(u) \wedge t(r) \langle t(r_{u,x})|} \tag{7}$$

In Eq. 7, the denominator corresponds to the number of ratings that have been entered by user u prior to rating $r_{u,x}$, i.e. the rating for which the dynamic average $DA_{previous}(r_{u,x})$ is calculated. This approach is based on the rationale that all past behaviour of the user is equally important in estimating her rating practices.

4   Performance Evaluation

In this section, we report on our experiments through which we compared the proposed algorithm with the $DA_{vicinity}$ and $DA_{previous}$ algorithms presented in [33], as well as the plain CF algorithm. We decided to consider in our evaluation both algorithms presented in [33] due to the following reasons:

1. the evaluation presented in [33] targets only the user-user CF approach, while in this paper we examine both the user-user and item-item CF approaches; hence, both the $DA_{vicinity}$ and the $DA_{previous}$ algorithms should be tested in order to evaluate their effectiveness in the item-item CF approach.
2. the comparison performed in [33] did not designate a clear winner between the $DA_{vicinity}$ and the $DA_{previous}$ algorithms; even though the $DA_{previous}$ algorithm outperforms the $DA_{vicinity}$ algorithm regarding prediction accuracy in most cases, there is a tie between the algorithms when they are applied in the MovieLens 100K dataset, while in the Netflix dataset the $DA_{vicinity}$ algorithm has been found to produce more accurate recommendations than the $DA_{previous}$ algorithm.

In this comparison we consider the following aspects:

1. prediction accuracy; for this comparison, we used two well-established error metrics, namely the mean absolute error (MAE) metric, as well as the Root Mean Squared Error (RMSE) that 'punishes' big mistakes more severely. RMSE was used in the Netflix competition [11],
2. the coverage of the algorithm, i.e. the percentage of the cases for which a prediction can be computed and
3. the probability that an algorithm computes the correct user rating. Since user ratings are typically integer numbers, while predictions are calculated as real numbers, for comparing the prediction to the actual user rating we round the prediction to the nearest integer. This is analogous to the practice used in the Netflix Competition [11].

To compute the MAE, the RMSE and the probability to compute the correct prediction, we employed the following techniques:

1. the standard "hide one" technique [30], which is extensively used in recommender systems research; each time, we hid a random rating in the database and then predicted its value based on the ratings of other non-hidden items. For each user, this procedure was executed for 10 randomly selected ratings entered by that particular user.
2. each time, we hid the last rating only from each user, and then predicted its value based on the ratings of other non-hidden items. One prediction for each user was formulated.
3. dropping the last rating from every user, and then applying the technique listed in item 2 above in the remaining dataset.

In all cases, the computation of the MAE, the RMSE and the correct prediction probability was performed considering all users in the database. All results were in close agreement (MAE: $\pm.0.005$; RMSE: $\pm.0.008$; correct predictions: $\pm.0.2\%$; % coverage: $\pm.0.4\%$), therefore in the rest of the paper we present only the results obtained from the standard "hide one" technique.

Regarding the algorithms presented in the related work section, except for those presented in [33], these are not directly comparable to our approach because they are designed to handle different phenomena, and more specifically concept drift (i.e. the change in users' interests), significance decay of old ratings and session identification. Nevertheless, to provide some insight on the magnitude of the improvement that can be achieved by different approaches that take into account temporal dynamics, we compare the performance of the DA_{next} algorithm against the following algorithms sourced from the literature:

1. from the category of forgetting algorithms, i.e. algorithms that decay the importance of old-aged ratings, we compare DA_{next} with the work of Vaz et al. [16];
2. from the set of algorithms targeting interest shift detection and exploitation, we compare DA_{next} against the ISCF algorithm [53];
3. from the domain of temporally-aware session-based algorithms, we compare DA_{next} with the work of Dror et al. [2], which examines the influence of the drifting effect in short-lived music listening sessions; and
4. from the category of temporally-aware matrix factorization algorithms, we compare DA_{next} against the algorithm proposed by Lo et al. [52], which tracks and exploits concept drift in each individual user latent vector.

The algorithms employing dynamic averages may exhibit different coverage, since the introduction of dynamic averages modifies the user-to-user and item-to-item similarity metrics, and henceforth users or items that are deemed "similar" when using the plain CF algorithm (i.e. when their standard Pearson or Adjusted Cosine similarity surpasses a threshold) may be deemed "not similar" when using the dynamic average-aware Pearson similarity or Adjusted Cosine Similarity, or vice versa. Under this condition, some users that are characterized as "grey sheep" [6] when using the plain CF algorithm (i.e. do not have enough near neighbours for a recommendation to be computed) may gain enough neighbours when using a dynamic average-based algorithm, thus increasing coverage; conversely some users for which a recommendation was computed using the plain CF algorithm may become "grey sheep" when using a dynamic average-based algorithm, in which case coverage decreases. An analogous phenomenon also appears in an item-item CF implementation.

For our experiments we used a machine equipped with six Intel Xeon E7 - 4830 @ 2.13 GHz CPUs, 256 GB of RAM and one 900 GB HDD with a transfer rate of 200 MBps, which hosted the datasets and ran the recommendation algorithms.

In the following paragraphs, we report on our experiments regarding ten datasets. Five of these datasets are obtained from Amazon [7, 8], three from MovieLens [9, 10] one from Netflix [11], while the last dataset is sourced from Ciao, a product review site, where users can post their experiences with products or services (the site, dvd.ciao.co. uk, has ceased its operations, however the datasets crawled from it still exist and are used in CF research). These ten datasets used in our experiments (a) contain reliable timestamps (most of the ratings within each dataset have been entered in real rating time and not in a batch mode), (b) are up to date (published between 1998 and 2016),

(c) are widely used as benchmarking datasets in CF research and (d) vary with respect to type of dataset (movies, music, books, videogames and automotive) and size (from 1 MB to 4.7 GB). The basic properties of these datasets are summarized in Table 1.

In each dataset, users initially having less than 10 ratings were dropped, since users with few ratings are known to exhibit low accuracy in predictions computed for them [26]. This procedure did not affect the MovieLens and the NetFlix datasets, because these datasets contain only users that have rated 20 items or more. Furthermore, we detected cases where for a particular user all her ratings' timestamps were almost identical (i.e. the difference between the minimum and maximum timestamp was less than 5 s). These users were dropped as well, since this timestamp distribution indicated that the ratings were entered in a batch mode, hence the assigned timestamps are not representative of the actual time that these ratings were given by the users.

Table 1. Datasets summary

Dataset name	#users	#ratings	#items	Avg. #ratings/user	DB size (in text format)
Amazon "Video-games" [7, 8]	8.1K	157K	50K	19.6	3.8 MB
Amazon "CDs and Vinyl" [7, 8]	41K	1,300K	486K	31.5	32 MB
Amazon "Movies and TV" [7, 8]	46K	1,300K	134K	29.0	31 MB
Amazon "Books" [7, 8]	295K	8,700K	2,330K	29.4	227 MB
Amazon "Automotive" [7, 8]	7.3K	113K	65K	15.5	2.6 MB
MovieLens "Old 100K Dataset" [9, 10]	0.94K	100K	1.68K	106.0	2.04 MB
MovieLens "Latest-20M","recommended for new research" [9, 10]	138K	20,000K	27K	145	486 MB
MovieLens "Latest 100K", "Recommended for education and development" (small) [9, 10]	0.7K	100K	9K	143	2.19 MB
NetFlix competition [11]	480K	96,000K	17.7K	200	4,700 MB
Ciao [51]	1.1K	40K	16K	36.3	1 MB

In the following paragraphs, we report on our findings regarding the experiments described above, using both a user-user CF implementation, which employs the standard Pearson correlation coefficient for measuring user similarity, and an item-item CF implementation, which employs the Adjusted Cosine Similarity for measuring item similarity.

4.1   The Amazon "Videogames" Dataset

The results obtained from the Amazon "Videogames" dataset, are depicted in Table 2. Column "% coverage" corresponds to the percentage of cases for which the algorithm could compute predictions, or –equivalently– when the number of near neighbors computed using the algorithm's similarity metric was adequate [28] to formulate a rating prediction.

We can observe that under both CF implementations (user-user and item-item), the DA_{next} algorithm achieves the best results regarding prediction accuracy. In more detail, under the user-user CF implementation, the DA_{next} algorithm achieves an improvement in MAE of 1.6% against the runner up, which is the $DA_{previous}$ algorithm and a 7.3% improvement against the plain CF algorithm. Considering the RMSE metric, the respective improvements are 1.3% and 5.7%. The DA_{next} algorithm also achieves the highest percentage of correct predictions, with its performance edge on this metric ranging from 0.5% to 0.7%. These improvements are achieved at the expense of a coverage drop of 2.2% against the plain CF algorithm, which is deemed to be tolerable; it is notable, however, that the DA_{next} algorithm achieves a better coverage percentage than the $DA_{previous}$ algorithm, which was the winner of the corresponding test in [33].

Table 2. Amazon "Videogames" dataset results

Method	User-user CF (Pearson correlation)				Item-item CF (adjusted cosine similarity)			
	MAE (out of 4)	RMSE	% correct predictions	% coverage	MAE (out of 4)	RMSE	% correct predictions	% coverage
Plain CF	0.777	1.082	32.04	**72.14**	0.383	0.596	66.61	**94.41**
DA$_{vicinity}$	0.752	1.048	32.15	70.86	0.377	0.582	66.89	93.96
DA$_{previous}$	0.732	1.033	32.26	69.76	0.375	0.579	67.22	93.97
DA$_{next}$	**0.720**	**1.020**	**32.76**	69.95	**0.368**	**0.561**	**67.86**	93.85

Regarding the item-item CF implementation, the DA_{next} algorithm has a performance edge of 1.9% on the MAE metric against the $DA_{previous}$ algorithm, which is the runner up, while the relevant improvement against the plain CF algorithm is 3.9%. Considering the RMSE metric, the respective improvements are 3.1% and 5.9%. With respect to the correct predictions metric, the DA_{next} algorithm is ranked first, having a performance lead of 0.6% against the $DA_{previous}$ algorithm which is ranked second, and a 1.2% performance lead compared to the plain CF algorithm. With respect to the coverage metric, the performance of the DA_{next} algorithm is almost equal to the other two dynamic averages approaches, lagging behind them by 0.1%, while the coverage deterioration of the DA_{next} algorithm as compared to the plain CF algorithm is 0.6%, which is deemed to be tolerable.

Overall, in this dataset the DA_{next} algorithm achieves considerable gains in rating prediction accuracy, under both the user-user and item-item CF implementations, at the expense of small to tolerable deteriorations in coverage.

4.2 The Amazon "CDs and Vinyl" Dataset

Table 3 illustrates the results obtained from the Amazon "CDs and Vinyl" dataset. In this dataset, the user-user CF implementation could formulate a prediction for 59.3% of the cases, while for the item-item CF implementation coverage increases to 86.6%.

Table 3. Amazon "CDs & Vinyl" dataset results

Method	User-user CF (Pearson correlation)				Item-item CF (adjusted cosine similarity)			
	MAE (out of 4)	RMSE	% correct predictions	% coverage	MAE (out of 4)	RMSE	% correct predictions	% coverage
Plain CF	0.702	1.010	29.15	**59.30**	0.335	0.557	64.87	**86.55**
DA$_{vicinity}$	0.682	0.984	29.36	58.66	0.332	0.544	65.94	86.27
DA$_{previous}$	0.669	0.969	29.28	57.71	0.331	0.539	66.73	86.19
DA$_{next}$	**0.663**	**0.957**	**29.74**	57.95	**0.324**	**0.525**	**67.32**	86.21

Regarding the rating prediction quality, the DA_{next} algorithm again outperforms all other algorithms, under both the user-user and the item-item CF implementations. In more detail, when considering the user-user CF implementation the DA_{next} algorithm achieves the lowest MAE metric (0.663), having a performance lead of 0.9% against the $DA_{previous}$ algorithm, which is ranked second, and a performance lead of 5.6% compared to the plain CF algorithm. For the RMSE metric, the respective performance edges are 1.2% and 5.2%. The DA_{next} algorithm is also ranked first with regards to the percentage of correct predictions metric, with its performance lead ranging from 0.4% over the performance of the $DA_{vicinity}$ algorithm which is ranked second regarding this metric, to 0.6% as compared to the plain CF algorithm. These benefits are achieved at the expense of a coverage drop of 1.3% as compared to the plain CF algorithm, which is considered to be tolerable. It is worth noting that coverage-wise, the DA_{next} algorithm achieves a slightly better performance than the $DA_{previous}$ algorithm, which is the runner up with respect to prediction accuracy.

With respect to the item-item CF implementation, the DA_{next} algorithm attains the lowest value for the MAE metric, which is 2.1% better than the MAE of the runner up algorithm ($DA_{previous}$) and 3.3% better than the value of the plain CF algorithm. Considering the RMSE metric, the respective improvements are 2.6% and 5.7%. The DA_{next} algorithm also computes the highest percentage of correct predictions, outperforming the DA_{next} algorithm, which is ranked second, by 0.6% and the plain CF algorithm by 2.5%. The DA_{next} algorithm exhibits a coverage drop of 0.3% against the plain CF algorithm, which is very small, while coverage-wise it is almost equivalent to the other two dynamic average-based algorithms.

Overall, in this dataset the DA_{next} algorithm achieves considerable gains in rating prediction accuracy, under both the user-user and item-item CF implementations, at the expense of very small to small deteriorations in coverage.

4.3   The Amazon "Movies & TV" Dataset

Table 4 illustrates the results obtained from the Amazon "Movies & TV" dataset. We can observe that, again, the proposed dynamic average-based algorithm DA_{next} achieves the best results under both user-user and item-item CF implementations. Considering user-user CF, the DA_{next} algorithm reduces the MAE by 7.0% as compared to the plain CF algorithm, while it also achieves a MAE reduction of 1.3%, compared to the $DA_{previous}$ algorithm which is the runner up. The respective improvements for the RMSE metric are 5.2% and 0.4%. The DA_{next} algorithm is also ranked first regarding the percentage of correct predictions, with its performance edge ranging from 0.5% (against

$DA_{previous}$) to 1% (against plain CF). The coverage of the DA_{next} algorithm, however, lags behind that of the plain CF algorithm by 1.6%, a drop which is deemed tolerable.

Considering the item-item CF implementation, the DA_{next} algorithm decreases the MAE by 3.1% against plain CF and by 1.8% against the runner up, which is the $DA_{previous}$ algorithm. The respective reductions in RMSE are higher (4.5% against the plain CF algorithm and 2.8% against the $DA_{previous}$ algorithm), indicating that the DA_{next} algorithm manages to correct some predictions with high errors (recall that the RMSE metric penalizes predictions with high errors). The DA_{next} algorithm is ranked first with respect to the correct prediction percentage by a margin ranging from 0.5% (against the $DA_{previous}$ algorithm) to 0.8% (against the plain CF algorithm). Finally, the coverage drop inflicted by the use of the DA_{next} algorithm is 0.6% as compared to the plain CF algorithm, which is deemed small.

Overall, in this dataset the DA_{next} algorithm achieves noteworthy improvements in rating prediction accuracy, under both the user-user and item-item CF implementations, while the losses in coverage imposed by the algorithm are rated from small to tolerable.

Table 4. Amazon "Movies & TV" dataset results

Method	User-user CF (Pearson correlation)				Item-item CF (adjusted cosine similarity)			
	MAE (out of 4)	RMSE	% correct predictions	% coverage	MAE (out of 4)	RMSE	% correct predictions	% coverage
Plain CF	0.738	1.046	37.00	**78.50**	0.393	0.617	66.74	**95.90**
$DA_{vicinity}$	0.713	1.014	37.37	77.26	0.390	0.611	66.81	95.60
$DA_{previous}$	0.695	0.996	37.55	76.91	0.388	0.606	67.09	95.30
DA_{next}	**0.686**	**0.992**	**38.01**	76.90	**0.381**	**0.589**	**67.56**	95.35

4.4   The Amazon "Books" Dataset

Table 5 illustrates the results obtained from the Amazon "Books" dataset. For the user-user CF, we can observe that again the DA_{next} algorithm is ranked first regarding prediction accuracy, its MAE being 2.6% less than the MAE of the plain CF algorithm and 0.5% smaller than the MAE of the $DA_{previous}$ algorithm, which is ranked second. The improvements regarding the RMSE metric are very similar to those of the MAE (2.5% and 0.7% respectively). The DA_{next} algorithm is also ranked first regarding the correct predictions percentage, by a narrow margin that ranges from 0.2% to 0.6%. The coverage achieved by the DA_{next} algorithm is 1% inferior to that achieved by the plain CF and the $DA_{vicinity}$ algorithms, which are tied for the first place; this drop, however, is deemed small to tolerable.

Table 5. Amazon "Books" dataset results

Method	User-user CF (Pearson correlation)				Item-item CF (adjusted cosine similarity)			
	MAE (out of 4)	RMSE	% correct predictions	% coverage	MAE (out of 4)	RMSE	% correct predictions	% coverage
Plain CF	0.625	0.883	43.67	**53.75**	0.301	0.466	71.97	**90.87**
$DA_{vicinity}$	0.619	0.876	43.96	**53.75**	0.297	0.461	72.02	90.24
$DA_{previous}$	0.612	0.867	44.04	53.36	0.288	0.450	72.21	89.46
DA_{next}	**0.609**	**0.861**	**44.26**	52.73	**0.278**	**0.441**	**72.71**	89.77

Regarding the item-item CF implementation, the DA_{next} algorithm achieves more substantial improvements than in the user-user setting: it achieves a reduction of 7.6% in the MAE, as compared to the plain CF algorithm and a reduction of 3.5% in the MAE against the runner up, which is the $DA_{previous}$ algorithm. The respective improvements regarding the RMSE are 5.4% and 2%. The DA_{next} algorithm also exhibits the best performance regarding the correct predictions percentage, surpassing the $DA_{previous}$ algorithm by 0.5% and the plain CF algorithm by 0.7%. These gains are achieved at the expense of a coverage drop, which is quantified to 1.1% against the plain CF algorithm; notably however, the DA_{next} algorithm attains better coverage than the runner up algorithm in terms of performance ($DA_{previous}$), by a small margin of 0.3%.

Overall, in this dataset the DA_{next} algorithm achieves substantial gains in rating prediction accuracy, under both the user-user and item-item CF implementations, while the losses in coverage imposed by the algorithm are rated from small to tolerable.

4.5   The Amazon "Automotive" Dataset

Table 6 illustrates the results obtained from the Amazon "Automotive" dataset. For the user-user CF, we can observe that the DA_{next} algorithm is ranked first regarding prediction accuracy, reducing the MAE by 7.6% in comparison to the plain CF algorithm; the runner-up algorithm regarding the MAE metric is $DA_{previous}$, which achieves a MAE 4.0% smaller than the plain CF algorithm, lagging behind DA_{next} by 3.6%. The ranking is the same regarding the RMSE metric, with DA_{next} being the top-performing algorithm, achieving an improvement in the RMSE by 8.0% in comparison to the plain CF algorithm, which surpasses the performance of $DA_{previous}$ –which is ranked second– by 3.8%. The DA_{next} algorithm is also ranked first regarding the correct predictions percentage, by a margin that ranges from 2.7% (against the $DA_{previous}$ algorithm) to 5.4% (against the $DA_{vicinity}$ algorithm). The coverage achieved by the DA_{next} algorithm is 1.4% inferior to that achieved by the plain CF algorithm, which is ranked first regarding this metric; this drop, however, is deemed small to tolerable.

Table 6. Amazon "Automotive" dataset results

Method	User-user CF (Pearson correlation)				Item-item CF (adjusted cosine similarity)			
	MAE (out of 4)	RMSE	% correct predictions	% coverage	MAE (out of 4)	RMSE	% correct predictions	% coverage
Plain CF	0.645	1.022	56.12	**53.16**	0.310	0.582	60.34	**91.77**
$DA_{vicinity}$	0.626	1.015	54.25	52.99	0.302	0.581	61.61	91.30
$DA_{previous}$	0.619	0.979	57.01	51.96	0.290	0.566	65.23	90.71
DA_{next}	**0.596**	**0.940**	**59.68**	51.75	**0.282**	**0.533**	**66.28**	90.37

Regarding the item-item CF implementation, the DA_{next} algorithm achieves higher performance improvements than in the user-user setting: it achieves a reduction of 9.0% in the MAE, as compared to the plain CF algorithm, with this improvement being better by 2.6%, as compared to that achieved by the runner up, which is the $DA_{previous}$ algorithm. The respective performance edges regarding the RMSE are 8.4% and 5.7%. The DA_{next} algorithm also attains the best performance regarding the correct predictions percentage, surpassing the $DA_{previous}$ algorithm by 1.1% and the plain CF algorithm by 5.9%. These gains are achieved at the expense of a coverage drop, which is quantified to 1.4% against the plain CF algorithm.

Overall, in this dataset the DA_{next} algorithm achieves substantial gains in rating prediction accuracy, under both the user-user and item-item CF implementations, while the losses in coverage imposed by the algorithm are rated from small to tolerable.

4.6   The MovieLens "Old 100K" Dataset

Table 7 depicts the results obtained from the MovieLens "Old 100K" dataset. In this dataset, which is a relatively dense one, we can observe that in both the user-user and the item-item CF implementations, practically no coverage drop is incurred by the introduction of the dynamic average-based algorithms, and coverage is close to 100% in all cases, with negligible variations. As shown in the next subsections, this behavior is consistent across all dense datasets (i.e. all MovieLens datasets and the Netflix dataset).

Regarding rating prediction quality, under the user-user CF implementation the DA_{next} algorithm has a marginal performance edge over the runner up, $DA_{previous}$, since it exhibits smaller values for both the MAE and the RMSE metric by 0.4%. In comparison to the plain CF algorithm the performance lead of the DA_{next} algorithm is considerably higher (MAE: 3.9%; RMSE: 3.5%). With respect to the correct predictions percentage criterion, the DA_{next} algorithm surpasses the performance of all other algorithms, having a lead of 1.1% against the $DA_{vicinity}$ algorithm which is ranked second and a lead of 2.2% against the plain CF algorithm.

Table 7. MovieLens "Old 100K" dataset results

Method	User-user CF (Pearson correlation)				Item-Item CF (adjusted cosine similarity)			
	MAE (out of 4)	RMSE	% correct predictions	% coverage	MAE (out of 4)	RMSE	% correct predictions	% coverage
Plain CF	0.735	0.939	42.34	99.82	0.622	0.797	48.99	99.90
$DA_{vicinity}$	0.715	0.916	43.38	99.83	0.618	0.790	49.37	99.90
$DA_{previous}$	0.709	0.910	43.32	**99.84**	0.611	0.778	50.22	99.90
DA_{next}	**0.706**	**0.906**	**44.49**	99.81	**0.603**	**0.760**	**50.82**	99.90

Considering the item-item CF implementation, the DA_{next} algorithm is again ranked first in all accuracy-related metrics. Regarding the MAE, the DA_{next} algorithm outscores the runner up (which is the $DA_{previous}$ algorithm) by 1.3% and the plain CF algorithm by 3.1%; in relation to the RMSE, the performance edge of the DA_{next} algorithm against the $DA_{previous}$ and the plain CF algorithms is 2.3% and 4.6%, respectively. Finally, the DA_{next} algorithm produces correct results in 0.6% more cases than the $DA_{previous}$ algorithm does, and in 1.8% more cases than the plain CF algorithm does.

Overall, in this dataset the DA_{next} algorithm achieves considerable gains in rating prediction accuracy, under both the user-user and item-item CF implementations, while practically no loss in coverage is sustained.

4.7   The MovieLens "Latest-20M - Recommended for New Research" Dataset

Table 8 depicts the results obtained from the MovieLens "Latest-20M, Recommended for New Research" dataset. As noted in the previous subsection, the coverage in this dataset is near 100% under both CF implementation strategies and remains practically unaltered by the introduction of dynamic average-based algorithms.

With respect to rating prediction quality, under the user-user CF implementation the DA_{next} algorithm is again ranked first, improving the MAE by 3.6% as compared to the plain CF algorithm and by 1.2% as compared to the $DA_{previous}$ algorithm, which is ranked second. The respective improvements considering the RMSE metric are 4.8% and 1.6%, respectively. Finally, the DA_{next} algorithm formulates the most correct predictions, surpassing the performance of the $DA_{previous}$ algorithm by 0.4% and that of the plain CF algorithm by 1.8%.

Table 8. MovieLens "Latest-20M - recommended for New Research" dataset results

Method	User-user CF (Pearson correlation)				Item-item CF (adjusted cosine similarity)			
	MAE (out of 9)	RMSE	% correct predictions	% coverage	MAE (out of 9)	RMSE	% correct predictions	% coverage
Plain CF	1.352	1.771	24.26	**99.96**	1.548	1.984	20.50	**99.99**
$DA_{vicinity}$	1.326	1.740	24.98	99.90	1.512	1.921	22.87	99.97
$DA_{previous}$	1.319	1.714	25.60	99.96	1.484	1.853	24.81	99.96
DA_{next}	**1.303**	**1.686**	**26.02**	99.94	**1.429**	**1.795**	**25.87**	99.96

Considering the item-item CF implementation, the DA_{next} algorithm has been found to produce the most accurate recommendations, exhibiting improvements in the MAE that range from 3.7% (against the $DA_{previous}$ algorithm) to 7.7% (against the plain CF algorithm); the corresponding improvements in RMSE range from 3.1% (against the $DA_{previous}$ algorithm) to 9.5% (against the plain CF algorithm). Finally, the DA_{next} algorithm produces the most correct predictions, having a performance lead of 1.1% in comparison to the $DA_{previous}$ algorithm, and a lead of 5.4% against the plain CF algorithm.

Overall, in this dataset the DA_{next} algorithm achieves considerable gains in rating prediction accuracy under both the user-user CF implementation and more substantial gains under the item-item CF implementation, while practically no loss in coverage is sustained.

4.8   The MovieLens "Latest 100K - Recommended for Education and Development" Dataset

Table 9 depicts the results obtained from the MovieLens "Latest 100K- Recommended for education and development" dataset. We can again notice that no coverage loss is introduced by the dynamic average-based algorithms, with coverage being near-100% in all cases.

With respect to rating prediction quality, under the user-user CF implementation scenario the DA_{next} algorithm is ranked first, achieving a reduction in the MAE of 3.2% in comparison to the plain CF and a reduction of 0.5% in comparison to the $DA_{previous}$ algorithm, which is ranked second. The respective gains for the RMSE metric are 4.3% and 1%, being higher than those of the MAE metric, indicating that the DA_{next} algorithm improves predictions with high errors. Finally, the DA_{next} algorithm computes approximately 1.7% more correct predictions than both the plain CF and $DA_{previous}$ algorithms, while it also exceeds the performance of the $DA_{vicinity}$ algorithm (which is the runner up for this metric) by 1.4%.

With regards to the item-item CF implementation scenario, the DA_{next} algorithm is again ranked first, achieving a 6.2% reduction in the MAE and 5.2% reduction in the RMSE, as compared to the plain CF algorithm. The $DA_{previous}$ algorithm is ranked second, lagging behind the DA_{next} algorithm by 2.0% regarding both the MAE and the RMSE metrics. Finally, the DA_{next} algorithm manages to produce the most correct predictions, computing 5.1% more correct predictions than the plain CF and 1.5% more correct predictions than the $DA_{previous}$ algorithm, which is ranked second.

Table 9. MovieLens "Latest 100K - recommended for education and development" dataset results

Method	User-user CF (Pearson correlation)				Item-item CF (adjusted cosine similarity)			
	MAE (out of 9)	RMSE	% correct predictions	% coverage	MAE (out of 9)	RMSE	% correct predictions	% coverage
Plain CF	1.404	1.858	24.16	99.57	0.999	1.374	34.34	**99.70**
$DA_{vicinity}$	1.376	1.816	24.44	**99.60**	0.987	1.368	35.08	99.68
$DA_{previous}$	1.366	1.796	24.07	99.40	0.956	1.330	37.99	99.65
DA_{next}	**1.359**	**1.778**	**25.83**	99.49	**0.937**	**1.303**	**39.46**	99.69

Overall, in this dataset the DA_{next} algorithm achieves considerable gains in rating prediction accuracy under the user-user CF implementation scenario and more substantial gains under item-item CF; in terms of coverage, practically no loss in coverage is sustained.

4.9 The "Netflix Competition" Dataset

Table 10 depicts the results obtained from the "Netflix Competition" dataset. Again, the coverage is close to 100% in all cases, and the losses sustained by the introduction of the dynamic average-based algorithms are negligible.

Regarding rating prediction quality, under the user-user CF scenario the DA_{next} algorithm is ranked first; it achieves a MAE improved by 3.7% against the plain CF algorithm and by 1.5% against the $DA_{vicinity}$ algorithm, which is ranked second; for the RMSE metric, the respective improvements are 4.4% and 1.4%. Finally, the DA_{next} algorithm computes 2.2% more correct predictions that the plain CF algorithm and 1.4% more correct predictions than the $DA_{vicinity}$ algorithm, which is ranked second for this metric.

Table 10. "Netflix Competition" dataset results

Method	User-user CF (Pearson correlation)				Item-item CF (adjusted cosine similarity)			
	MAE (out of 4)	RMSE	% correct predictions	% coverage	MAE (out of 4)	RMSE	% correct predictions	% coverage
Plain CF	0.758	0.960	41.42	**99.10**	0.857	1.061	33.61	**99.40**
$DA_{vicinity}$	0.741	0.931	42.25	99.03	0.811	0.956	39.98	99.30
$DA_{previous}$	0.752	0.936	42.12	99.00	0.808	0.954	40.08	99.23
DA_{next}	**0.730**	**0.918**	**43.60**	99.02	**0.761**	**0.892**	**44.30**	99.25

Considering the item-item CF implementation, the DA_{next} algorithm outperforms all other algorithms by a wider margin. For the MAE criterion, it achieves an improvement of 11.2% against the plain CF algorithm, and 5.8% against the $DA_{previous}$ algorithm, which is ranked second; the respective improvements for the RMSE metric are 15.9% and 6.5%. Finally, the DA_{next} algorithm computes the most correct predictions, having a performance edge of 4.2% against the $DA_{previous}$ algorithm, which is the runner up, and an edge of 10.7% against the plain CF algorithm.

Overall, in this dataset the DA_{next} algorithm achieves considerable gains in rating prediction accuracy under the user-user CF implementation and more substantial gains under the item-item CF implementation, while practically no loss in coverage is sustained.

4.10 The "Ciao" Dataset

Table 11 illustrates the results obtained from the "Ciao" dataset. For the user-user CF, we can observe that the DA_{next} algorithm is ranked first regarding prediction accuracy, reducing the MAE by 6.1% in comparison to the plain CF algorithm; the runner-up algorithm regarding the MAE metric is $DA_{previous}$, which achieves a MAE 3.4% smaller than the plain CF algorithm, lagging thus behind DA_{next} by 2.7%. The ranking is the same regarding the RMSE metric, with DA_{next} being the top-performing algorithm, achieving an improvement in the RMSE by 9.7% against the plain CF algorithm, which surpasses the performance of $DA_{previous}$ –which is ranked second– by 3.0%. The DA_{next}

algorithm is also ranked first regarding the correct predictions percentage, by a margin that ranges from 1.7% (against the $DA_{previous}$ algorithm) to 3.2% (against the plain CF algorithm). The coverage achieved by the DA_{next} algorithm is 1.4% inferior to that achieved by the plain CF algorithm, which is ranked first regarding this metric; this drop, however, is deemed small to tolerable.

Table 11. "Ciao" dataset results

Method	User-user CF (Pearson correlation)				Item-item CF (adjusted cosine similarity)			
	MAE (out of 4)	RMSE	% correct predictions	% coverage	MAE (out of 4)	RMSE	% correct predictions	% coverage
Plain CF	0.853	1.089	37.40	**76.86**	0.378	0.567	68.06	**99.26**
DA$_{vicinity}$	0.844	1.064	37.86	76.21	0.373	0.560	69.01	**99.26**
DA$_{previous}$	0.824	1.016	38.93	75.96	0.366	0.550	69.99	99.13
DA$_{next}$	**0.801**	**0.983**	**40.64**	75.47	**0.359**	**0.534**	**71.60**	99.11

Regarding the item-item CF implementation, the DA_{next} algorithm achieves a reduction of 5.0% in the MAE, as compared to the plain CF algorithm, with this improvement being better by 1.9%, as compared to that achieved by the runner up, which is the $DA_{previous}$ algorithm. The respective performance edges regarding the RMSE are 5.8% and 2.8%. The DA_{next} algorithm also attains the best performance regarding the correct predictions percentage, surpassing the $DA_{previous}$ algorithm by 1.6% and the plain CF algorithm by 3.5%. These gains are achieved at the expense of a practically negligible coverage drop, which is quantified to 0.15%.

Overall, in this dataset the DA_{next} algorithm achieves substantial gains in rating prediction accuracy, under both the user-user and item-item CF implementations, while the losses in coverage imposed by the algorithm are rated from negligible to tolerable.

4.11   Algorithms Comparison

In this subsection, we consolidate our findings from all datasets, to provide a comprehensive overview of the algorithms' performance regarding the prediction accuracy metrics. In all comparisons, the performance of the plain CF algorithm is taken as a baseline. We also compare the proposed algorithm against other approaches that have been published and evaluated in the literature.

Figure 1 depicts the improvement in the MAE achieved by all dynamic average-based algorithms under the user-user CF implementation scenario. Clearly, the DA_{next} algorithm achieves the best results, with its performance lead being on average approximately 1.5% against the $DA_{previous}$ algorithm which is the runner up; the respective reduction in the MAE against the baseline algorithm is 5.1% on average. It is worth noting that the DA_{next} algorithm surpasses the performance of both other algorithms in all datasets, while the $DA_{previous}$ algorithm is ranked second in 9 datasets and third in one dataset (Netflix). From the detailed examination of our results, the DA_{next} algorithm formulated the prediction with the lowest error in 94.3% of the prediction formulation requests across all datasets (this percentage includes ties for the first place, i.e. cases where the DA_{next} algorithm and some other algorithm(s) produced the same prediction, and this was the closest prediction to the actual rating).

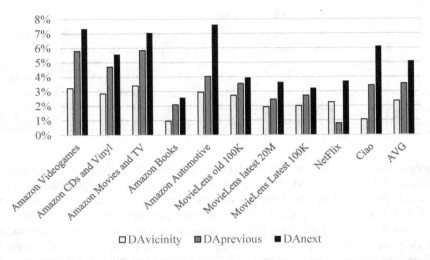

Fig. 1. MAE improvement achieved by the dynamic average-based algorithms under the user-user CF implementation

Figure 2 presents the respective improvements regarding the RMSE metric. In five datasets (and on average), the improvements are very similar to those of the MAE metric shown in Fig. 1, indicating that prediction improvements are spread uniformly among predictions with high and low errors. In three datasets (MovieLens latest 20M; MovieLens Latest 100K; and Ciao), the improvement in the RMSE metric is higher than the improvement in the MAE metric by a margin ranging from 1.1% to 3.6%, indicating that the DA_{next} algorithm manages to eliminate some high errors in predictions, while in two other datasets (Amazon Videogames and Amazon Movies and TV), the improvements in the MAE metric surpass those in the RMSE metric by 1.6% and 1.9% respectively, indicating that the DA_{next} algorithm mostly adjusts predictions with low errors. Again, the DA_{next} algorithm is consistently ranked first across all datasets, with its average performance lead against the runner up algorithm ($DA_{previous}$) being 1.5%, and the respective performance lead against the plain CF algorithm being 5.3%.

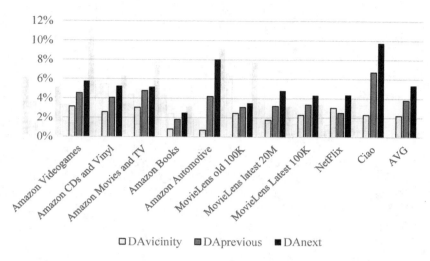

□ DAvicinity   ■ DAprevious   ■ DAnext

Fig. 2. RMSE improvement achieved by the dynamic average-based algorithms under the user-user CF implementation

Figure 3 illustrates the improvement in the MAE achieved by the dynamic average-based algorithms under the item-item CF implementation scenario. Again, the DA_{next} algorithm is consistently ranked first across all datasets, achieving a reduction of 6.0% on average against the baseline algorithm, and surpassing the performance of the runner up algorithm ($DA_{previous}$) by 2.6% on average. Performance gains in this case exhibit higher variations than those under the user-user implementation scenario, mainly owing to the Netflix dataset in which the DA_{next} algorithm achieves very substantial improvements in the MAE metric (11.2% against the plain CF algorithm and 5.5% against the runner up, which is the $DA_{previous}$ algorithm). From the detailed examination of our results, the DA_{next} algorithm formulated the prediction with the lowest error in 91.4% of the prediction formulation requests across all datasets.

Figure 4 presents the relevant improvements regarding the RMSE metric. Again, DA_{next} achieves the best results, with its performance lead against the $DA_{previous}$ algorithm, which is the runner up, being equal to 3.0% on average; the respective RMSE reduction against the baseline algorithm is 7.1% on average. In relation to the baseline algorithm, the average RMSE metric improvement is higher than that of the MAE metric by approximately 1.1%, indicating that the DA_{next} algorithm achieves to eliminate some high prediction errors. Considering individual datasets, the RMSE metric improvement is higher than the improvement of MAE in seven of the datasets (Amazon Videogames, Amazon CDs and Vinyl, Amazon Movies and TV, MovieLens old 100K, MovieLens latest 20M, Netflix and Ciao); in the remaining four three (Amazon Books, Amazon Automotive and MovieLens Latest 100K), the improvement in the RMSE metric lags behind the improvement of the MAE metric by a margin ranging from 0.6% to 2.2%, indicating that in these datasets the DA_{next} algorithm mostly improves predictions with low errors.

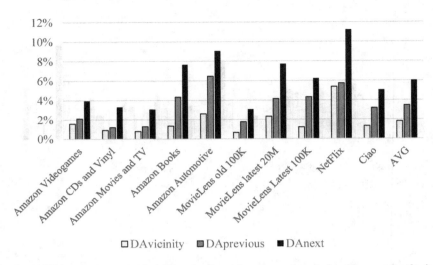

Fig. 3. MAE improvement achieved by the dynamic average-based algorithms under the item-item CF implementation

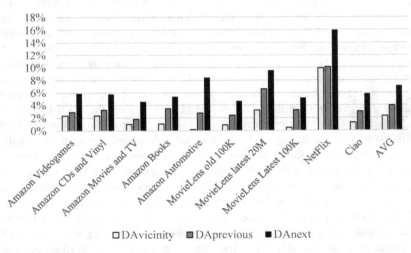

Fig. 4. RMSE improvement achieved by the dynamic average-based algorithms under the item-item CF implementation

Figure 5 illustrates the improvements regarding the correct prediction percentage for all dynamic average-based algorithms under the user-user CF implementation scenario. The DA_{next} algorithm achieves improvements ranging from 0.4% to 3.6% against the baseline algorithm. Regarding the dynamic average-based algorithms, we can observe that the DA_{next} algorithm is consistently ranked first across all datasets; the $DA_{previous}$ algorithm is ranked second in six of the datasets, lagging behind the DA_{next}

algorithm by a margin ranging between 0.22% and 2.67% (1.1% on average), while the $DA_{vicinity}$ algorithm is ranked second in the remaining four datasets, falling behind the performance of DA_{next} by a margin ranging between 0.3% and 5.43% (1.50% on average).

Figure 6 depicts the corresponding findings for the item-item CF implementation. Again, the DA_{next} algorithm is ranked first across all datasets, with the performance gains being considerably higher: the improvement against the baseline algorithm is 3.8% on average, ranging from 0.7% to 10.7%, while in comparison to the runner up, $DA_{previous}$, the performance edge of the DA_{next} algorithm is 1.2% on average, ranging from 0.5% to 4.2%. A significant part of this performance edge is owing to the results of the Netflix dataset, where the DA_{next} algorithm has the widest performance gap from the other algorithms. Besides the Netflix dataset, we can observe that the DA_{next} algorithm achieves its highest performance improvements in the latest MovieLens datasets (MovieLens latest 20M and MovieLens Latest 100K) and the Amazon "Automotive" dataset. The Netflix and both the Movielens datasets share the property of being denser than other datasets (and being the only dense datasets in the experiment), with their $\frac{\#ratings}{\#users * \#items}$ ratio exceeding 1%, while in the rest of the datasets this ratio ranges from 0.001% (Amazon Books) to 0.537% (MovieLens old 100K). However, further investigation is required to determine whether this behavior is owing solely to the density of the datasets, or to other properties as well. Interestingly, the DA_{next} algorithm achieves substantial improvements in the Amazon "Automotive" dataset too, which is relatively sparse.

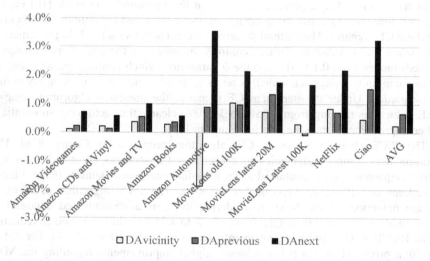

Fig. 5. Correct predictions percentage improvement achieved by the dynamic average-based algorithms under the user-user CF implementation

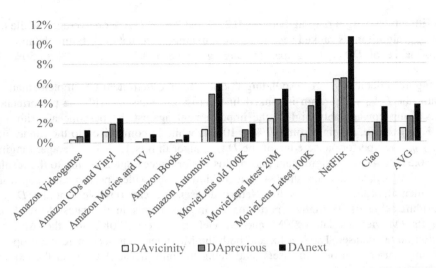

Fig. 6. Correct predictions percentage improvement achieved by the dynamic average-based algorithms under the item-item CF implementation

Summarizing, we can clearly see that in all datasets and under both CF implementation scenarios, the proposed algorithm outperforms all other algorithms achieving (1) the highest MAE reduction, (2) the highest RMSE reduction and (3) the highest correct predictions' percentage.

In relation to other approaches reported in the literature, Vaz et al. [16] exploit temporal dynamics by considering the age of user ratings and community ratings in an item-item CF scenario. The method presented therein achieves a MAE improvement of 0.2% when the a parameter (which controls the way that the age of user ratings is handled) ranges from 0.1 to 0.6 and the b parameter (which controls the way that the age of community ratings is handled) is set to 0, i.e. when the age of community ratings is disregarded. Under the item-item CF scenario, the presented algorithm achieves MAE reductions ranging from 3.05% to 11.20%, clearly thus achieving substantially higher improvements than the one presented in [16].

The ISCF (Interest Sequence CF) algorithm proposed by Cheng et al. [53] accommodates temporal dynamics in a user-user CF scenario, by considering user interest sequences, and is evaluated against the four real-world datasets (Ciao, Flixter, MovieLens old 100K, MovieLens latest 100K). The ISCF algorithm achieves an average reduction on the MAE (considering the four aforementioned datasets) by 2.41% (with improvements ranging from 0.21% to 4.33%), while the average reduction on the RMSE is 3.0% (with improvements ranging from 0.59% to 7.81%). The DA_{next} algorithm proposed in this paper achieves higher improvements regarding the MAE and the RMSE than ISCF on average, both in individual datasets and on overall average, as illustrated in Table 12.

Table 12. Accuracy improvements achieved by the DA_{next} and the ISCF [53] algorithms

	Ciao		MovieLens old 100K		MovieLens latest 100K		Average (over all tested datasets)	
	MAE	RMSE	MAE	RMSE	MAE	RMSE	MAE	RMSE
ISCF	0.21%	0.59%	3.23%	0.21%	0.59%	2.24%	2.41%	3.0%
DA_{next}	6.10%	9.73%	3.95%	3.51%	3.20%	4.3%	5.1%	5.3%

Dror et al. [2] propose an algorithm for identifying and exploiting drifts in short-lived music listening sessions. Their algorithm is evaluated against the Yahoo Music! dataset; the algorithm presented in [2] encompasses two steps that exploit temporal dynamics, namely the *user session bias* and the *items temporal dynamics bias*. These two steps achieve a cumulative improvement in RMSE equal to 3.98%; on the other hand, the proposed algorithm has been found to improve RMSE by 6.52% under the user-user scenario and by 4.89% under the item-item scenario. Therefore, in this case the DA_{next} algorithm is found to achieve higher improvement levels, under both the user-user and the item-item CF scenarios, than the provisions for exploitation of temporal dynamics proposed by Dror et al. [2].

Finally, Lo et al. [52] develop a temporal matrix factorization approach for tracking concept drift in each individual user latent vector. The method proposed therein is applied on four real-world datasets, achieving reductions in the RMSE metric ranging from 0.24% (when applied on the MovieLens 20M dataset) to 5.04% (when applied on the Ciao dataset); the average RMSE metric improvement achieved by the algorithm presented in [52] considering the four real-world datasets is 1.73%. The respective improvements regarding the RMSE metric achieved by the algorithm proposed in this paper are as follows: regarding the MovieLens 20M dataset, the RMSE is decreased by 4.8% under the user-user scenario and by 9.5% under the item-item scenario; in regards to the Ciao dataset, the RMSE is decreased by 9.7% under the user-user scenario and by 5.8% under the item-item scenario; finally, the average RMSE reduction achieved by the proposed algorithm across all ten examined datasets is 5.3% under the user-user scenario and 7.1% under the item-item scenario. Recapitulating, the algorithm proposed in this paper achieves more substantial improvements in rating prediction accuracy than the one proposed in [52], both considering individual datasets and average performance, and this performance edge is achieved under the user-user CF scenario as well as the item-item CF scenario.

4.12 Algorithm Complexity and Scalability

In this subsection, we investigate the complexity and the scalability of the proposed algorithm, and compare them with the complexity and scalability of the other algorithms examined in our experiments [33]. In our investigation, we consider all phases of the algorithms, i.e. (i) bootstrap (initial computations of dynamic averages and Pearson similarities), (ii) computation of recommendations and (iii) update of dynamic averages and Pearson similarities.

Bootstrap Phase. Within the bootstrap phase, the dynamic average-based algorithms (DA_{next}, $DA_{previous}$ and $DA_{vicinity}$) need in order to compute the dynamic averages for each rating of the users as well as the Pearson similarities between users. The plain CF algorithm needs to compute the global average of each user's ratings and the Pearson similarities between users. For each of the algorithms, the relevant complexities are presented in the following paragraphs.

Computation of Needed Averages. For the DA_{next} algorithm, Eq. 4 indicates that for each rating of a user all ratings subsequently entered by the same user need to be examined to compute the rating's dynamic average; under this view, the complexity of the calculation of the dynamic averages for each user is $O(ru^2)$, where ru is the number of ratings of the user. However, Listing 1 shows an optimization for the procedure of calculating the dynamic averages, according to which ratings are sorted in ascending timestamp order and then the sorted list is traversed from the end to the beginning, computing at each step the relevant rating's dynamic average by only considering the previous rating's value and the results of the computations made in the previous steps. Thus the complexity of computing the dynamic averages for each user under the DA_{next} algorithm is $O(ru * log(ru))$, i.e. the complexity of the sorting phase, which dominates the complexity of the whole operation. The complexity of computing the dynamic averages for all users is $O(N * \bar{r} * \log(\bar{r}))$, where N is the number of users and $\bar{r}$ is the average number of ratings per user.

Regarding $DA_{previous}$ algorithm, the same technique can be employed for computing dynamic averages (with the sorted list being traversed from the oldest rating to the newest one), hence the complexity of the computing the dynamic averages for all users is again $O(N * \bar{r} * \log(\bar{r}))$.

In the case of the temporal vicinity algorithm, $DA_{vicinity}$, Eq. 5 indicates that, when computing the dynamic average for a specific rating $r_{u,i}$, every rating $r_{u,i'}$ entered by the same user u is assigned a weight, based on its temporal vicinity to $r_{u,i}$ and subsequently its value is multiplied by that weight to compute the dynamic average of $r_{u,i}$. Therefore the complexity of calculating the dynamic average for a specific rating is $O(ru)$ and consequently the complexity of computing all dynamic averages for a specific user's ratings is $O(ru^2)$ and the complexity of calculating all users' dynamic averages is $O(N * \bar{r}^2)$. An optimization is possible in this procedure: when computing the dynamic average for a rating $r_{u,i}$ the initial and the final part of the temporally sorted rating list for which $w_{u,x}(r_{u,i}) < \varepsilon$, where ε is a small value (e.g. 10^{-2}), which will have a minimal impact at the computation of the dynamic average of $r_{u,i}$ could be excluded from the computation. However, due to the fact that the $DA_{vicinity}$ algorithm achieved the smallest improvements out of all the dynamic average algorithms considered, such optimizations were not considered further.

Finally, the plain CF algorithm computes each user's global average with a single pass along the user's ratings, therefore complexity of calculating of the global average for a user u is $O(ru)$, and the complexity of computing all users' global averages is $O(N * \bar{r})$.

Computation of Pearson Similarities. The method for computing the Pearson similarities between users is common to all four algorithms; the only difference between the plain CF algorithm on the one hand and the dynamic average-based algorithms on the other is that the plain CF algorithm uses the single global average per user for adjusting each rating $(\overline{R_X}$ and $\overline{R_Y}$, c.f. Eq. 1), while the dynamic average-based algorithms use a precomputed, rating-specific average $(DA_{Alg}(R_{X,i})$ and $DA_{Alg}(R_{Y,i})$, c.f. Eq. 2). This does not affect the complexity of the operations, since the same amount of computations takes place. In practice, performance differences may occur due to the fact that global averages may be stored in registers for fast access, while dynamic averages should be fetched from main memory, however this difference has been quantified to be small (measurements are presented below). In terms of complexity, for computing the Pearson similarity between two users X and Y the items rated in common by both users need to be identified and for each of these pair of item ratings simple computations are performed. The identification of items commonly rated among two users can be achieved by creating a hash set for one of the user's ratings and then iterating over the other user's ratings and examining whether they exist in the hash set. With a sufficiently large key space value for the hash function (which is always possible since users have at most a few thousand ratings and a hash set with tens of millions of distinct keys can be easily accommodated in memory), insertion and lookup in the hash set is O(1), therefore the complexity of the creation of the hash set is $O(r_x)$ and the complexity of the lookup is $O(r_Y)$, where r_X and r_Y is the number of ratings entered by users X and Y, respectively. Consequently, the complexity of the computation of the Pearson similarity between two users X and Y is $O(r_x) + O(r_Y)$, or in the general case $O(2 * \bar{r})$. Since the Pearson similarity between any pair of users needs to be computed, the overall complexity for the computation of the Pearson similarity is $O(2 * N^2 * \bar{r})$; since however the Pearson similarity is symmetric (i.e. $Pearson\_sim(X, Y) = Pearson\_sim(Y, X)$), the number of computations can be reduced to the half, yielding an overall complexity of $O(N^2 * \bar{r})$.

Table 13 summarizes the result of the complexity analysis for the bootstrap phase of the four algorithms. Note that in the case of the plain CF algorithm, the complexity of the Pearson similarity computation phase dominates the complexity of the dynamic average computation phase, hence in the overall complexity only the former appears.

Regarding the disk storage space needed, the plain CF algorithm needs to store only triples of the form *(user, item, rating)*, while all dynamic average-based algorithms need to extend the triple to accommodate the rating timestamp. In many cases, the timestamp is stored as seconds since the epoch (1/1/1970), for which 8 bytes are sufficient. Even with datasets with billions (10^9) of ratings, this extension can be accommodated, since even commodity hardware supports storages at the TB level (10^{12}).

Table 13. Complexity analysis for the algorithms; bootstrap phase

Method	Dynamic average computation complexity	Pearson similarity computation complexity	Overall complexity
Plain CF	$O(N * \bar{r})$	$O(N^2 * \bar{r})$	$O(N^2 * \bar{r})$
DA$_{vicinity}$	$O(N * \bar{r}^2)$	$O(N^2 * \bar{r})$	$O(N * \bar{r}^2) + O(N^2 * \bar{r})$
DA$_{previous}$	$O(N * \bar{r} * \log(\bar{r}))$	$O(N^2 * \bar{r})$	$O(N * \bar{r} * \log(\bar{r})) + O(N^2 * \bar{r})$
DA$_{next}$	$O(N * \bar{r} * \log(\bar{r}))$	$O(N^2 * \bar{r})$	$O(N * \bar{r} * \log(\bar{r})) + O(N^2 * \bar{r})$

Finally, in terms of memory storage, the dynamic average-based algorithms require the storage of the dynamic average along with each rating. Dynamic averages are represented with a float-typed value, requiring a four additional bytes. Taking into account that the memory capacity of contemporary high-end servers has substantially increased at the TB level (e.g. [48] can accommodate more than 6 TB of memory), this extension is not expected to be a problem. It is worth noting that the $DA_{previous}$ and the DA_{next} algorithms may totally drop the timestamps from a user's ratings after the user's dynamic averages have been computed, hence it is not necessary at any time point to accommodate both all ratings' timestamps and dynamic averages (the $DA_{vicinity}$ algorithm may need to retain the timestamp in memory, in order to recompute the dynamic averages when new ratings are entered).

In the case that the above quantified increases in storage space requirements is a consideration, storage space needs may decrease by computing dynamic averages per time window (e.g. week; month; year), in a fashion similar to the one described in [23]. The study of the effect that such an approach would have on the quality of formulated predictions is part of our future work.

Figure 7 illustrates the time needed to compute all ratings' (dynamic) averages for the various datasets under each of the four examined algorithms.

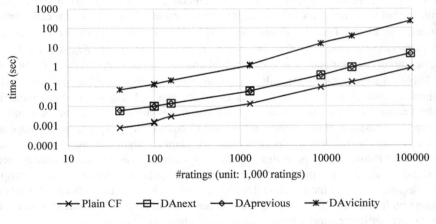

Fig. 7. Time needed for (dynamic) average computation

In Fig. 7 (n.b. both axis are in logarithmic scale) we can notice that the time needed by the plain CF algorithm to compute all global averages is less than 1 s (0.837 s. for the Netflix dataset which contains 100M ratings), while the respective time needed by the DA_{next} and $DA_{previous}$ algorithms (whose lines fully coincide) is always less than 5 s (4.879 s for the Netflix dataset); this increment is definitely considered manageable. On the other hand, the $DA_{vicinity}$ algorithm needs significantly more time, up to 230 s for the Netflix dataset.

Figure 8 illustrates the time needed to compute the Pearson similarities for all user pairs. The time needed by all algorithms has been found to be almost identical, with the dynamic average-based algorithms needing approximately 0.98% more time than the plain CF algorithm, which –as stated above– is attributed to the fact that global averages may be stored in registers for fast access, while dynamic averages must be fetched from main memory. While generally the time needed appears to scale linearly with the number of ratings, we can notice three data points where scaling is not linear:

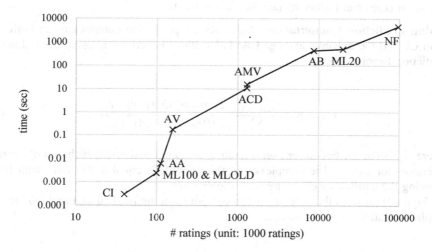

Fig. 8. Time needed for computing the pairwise Pearson similarities between users

1. The time needed for the Amazon Videogames dataset (denoted as AV in Fig. 8) is 72 times higher than the time needed for the MovieLens "Old 100K Dataset" (denoted as MLOLD), although the number of ratings is only 1.5 times higher. This is attributed to the fact that the number of users in the Amazon Videogames dataset is 8 times higher than the respective number of users in the MovieLens "Old 100K Dataset" and, as indicated by the complexity formulas in Table 13, the time needed is proportional to the square of the number of users, while scaling linearly with the average number of ratings.
2. The time needed for the Amazon Movies dataset (denoted as AMV in Fig. 8) is 1.3 times higher than the time needed for the Amazon CDs and Vinyl dataset (denoted as ACD), despite the fact that both datasets contain the same number of ratings. Again, this is attributed to the fact that the Amazon Movies dataset contains a higher number of users than the Amazon CDs and Vinyl dataset (approximately 11% higher).
3. Finally, the time needed for the Amazon Books dataset (denoted as AB) is almost equal to the time needed for the MovieLens "Latest 20M" dataset (denoted as ML20), although the latter contains 2.3 times more ratings than the former. This is

again owing to the fact that the Amazon Books dataset involves a significantly higher number of users than the MovieLens "Latest 20M" (295K users vs. 138K or 2.13 times more).

In all cases we can observe that the computation of all Pearson similarities requires from 0.3 ms (for the Ciao dataset, labeled as CI) to 75 min for a dataset containing 100M ratings (the Netflix dataset, labeled as NF), hence it is considered feasible. Since computation of Pearson similarities is clearly parallelizable (computations for any pair of users can proceed parallelly with the computation of any other pair), adding more execution cores can further reduce the time needed.

Rating Prediction Computation Phase. Rating prediction computation in both the plain CF and the dynamic average-based algorithms is performed using the following prediction formula [3]:

$$\widehat{r_{u,i}} = \overline{r_u} + \frac{\sum_{v \in NN(u)} Pearson\_sim(u, v) * r_{v,i}}{\sum_{v \in NN(u)} |Pearson\_sim(u, v)|} \tag{8}$$

where $\widehat{r_{u,i}}$ is the prediction for user u's rating for item i and $NN(u)$ is the set of nearest neighbors for user u. The complexity of this formula is equal to $O(|NN|)$, with $|NN|$ denoting the cardinality of the nearest neighbor set.

Figure 9 presents the time needed to compute a rating prediction, in relation to the number of ratings in the database.

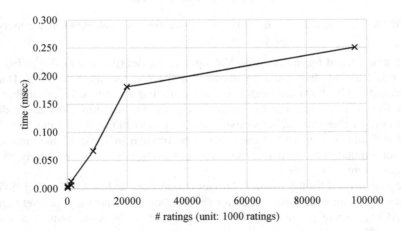

Fig. 9. Time needed for computing a rating prediction for a user

We can observe that the recommendation formulation time is under 1 ms in all cases. The time needed to compute a rating prediction increases with the number of ratings, owing to the fact that in our experiments the set of nearest neighbors was allowed to contain all users having a positive Pearson correlation with the user that the prediction was computed for. Pruning the nearest neighbor set to contain the top N similar users or users having a similarity above a higher threshold would render the computation even more efficient.

Database Update Phase. The ratings database is updated when users enter new ratings. When a rating is entered, the average(s) related to the user's ratings are modified, and additionally the user's Pearson similarity to other users changes. In the following subsections, we elaborate on the complexity and scalability of the database update phase, under the four algorithms considered in the evaluation section.

Updating a user's averages.
Under the plain CF algorithm, when a new rating r_{new} is entered in the database, the new global average can be computed using Eq. 9:

$$avg_{u,new} = \frac{avg_{u,current} * |ratings_u| + r_{new}}{|ratings_u| + 1} \tag{9}$$

where $ratings_u$ is the set of ratings that have been entered by user u before r_{new} was entered and $avg_{u,current}$ is the current global average of user u. Therefore, the complexity of updating a user's global average upon insertion of a new rating is $O(1)$.

When the $DA_{previous}$ algorithm is employed, the arrival of a new rating r_{new} entered by user u necessitates only the computation of the dynamic average for the newly entered rating: indeed, since the dynamic average of any rating r is based only on ratings entered by the same user *before* r, the dynamic averages of ratings previously entered by the same user are not affected. If r_{last} is the last rating entered by user u before r_{new}, then the dynamic average for r_{new} can be computed using Eq. 10:

$$DA_{previous}(r_{new}) = \frac{DA_{previous}(r_{last}) * |ratings_u| + r_{new}}{|ratings_u| + 1} \tag{10}$$

where $ratings_u$ is the set of ratings that have been entered by user u before r_{new} was entered. Consequently, the complexity of updating a user's dynamic averages upon insertion of a new rating is $O(1)$.

When the DA_{next} algorithm is used, the arrival of a new rating r_{new} entered by user u necessitates the computation of the dynamic averages for *all* ratings entered by u: this is due to the fact that the newly entered rating r_{new} has a greater timestamp than all existing ratings r entered by u, and therefore by virtue of Eq. 4 it affects the respective dynamic averages. To recalculate the dynamic averages of all ratings, the algorithm shown in Listing 1 can be employed. However, since the ratings are already ordered in ascending timestamp order, the sorting operation can be skipped, and therefore the complexity is reduced from $O(r_u * log(r_u))$ to $O(r_u)$, accounting for a single traversal of the ratings in descending timestamp order.

Under the $DA_{vicinity}$ algorithm, when a new rating r_{new} is entered by user u, the dynamic averages for *all* ratings entered by u must be calculated anew. This is because (a) since the newly entered rating will have a timestamp greater than all other ratings in the database, the denominator of Eq. 5 changes, and consequently the weight of every rating is modified and (b) the newly entered rating should be considered in the computation of the dynamic average of every rating entered by user u, according to Eq. 6. As shown in subsection "Bootstrap phase", the complexity of computing the dynamic averages for all ratings entered by some user u is $O(r_u^2)$, where r_u is the number of ratings entered by u.

Figure 10 illustrates the time needed for recomputing a user's dynamic averages in the context of the datasets used in this paper. We observe that the time needed for DA_{next} algorithm ranges between 0.54 and 10 μsec, while the time needed by the $DA_{previous}$ and the plain CF algorithm remains constant. Finally, the $DA_{vicinity}$ algorithm exhibits the worst performance among the examined algorithms.

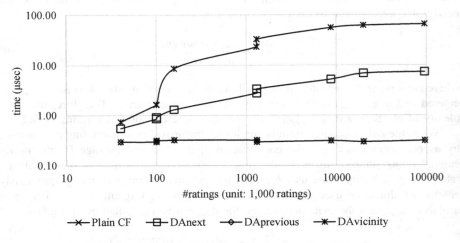

Fig. 10. Time needed for recomputing a user's average(s)

Recomputing a User's Pearson Similarities. When a new rating is entered, the Pearson similarities between a user and other users need to be recomputed. For the plain CF and the $DA_{previous}$ algorithm, where the global average and the dynamic averages respectively are not affected, this needs to be performed only for users that have rated the item referenced in the newly entered rating. For the DA_{next} and $DA_{vicinity}$ algorithms, similarities with all users need to be computed afresh, because the dynamic averages of all ratings change, and this affects the outcome of the dynamic average-aware Pearson similarity (c.f. Eq. 2).

Figure 11 depicts the performance of the procedure for recomputing the Pearson similarity between the user for which a new rating was entered and all other users in the dataset. For the Netflix dataset, which contains 480K users, the time needed is approximately 30 ms. It has to be noted that parallelism is not as efficient in this case as it has proven in the case of computing the pairwise similarities between all users, because the time needed to actually perform the calculations is now much smaller, therefore the overhead of thread creation is now a considerable portion in the overall execution time.

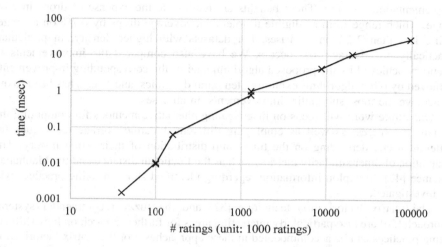

Fig. 11. Time needed to recompute the Pearson similarities between a user and all other users

Besides having a high computation cost, recomputing the Pearson similarities upon the arrival of each new rating is of low utility, because the results of some of these computations will be overwritten almost instantly, as new ratings by other users enter the database. For these reasons, the recomputation of the Pearson similarities can be performed in an offline fashion, either periodically or when a substantial number of new ratings has been amassed.

5   Conclusion and Future Work

In this paper we have introduced a novel dynamic average-based algorithm, DA_{next}, which is able to better follow the variations of user rating practices and, consequently, is able to produce more accurate rating predictions. The proposed algorithm has been experimentally verified using ten datasets and compared to other dynamic average-based algorithms presented in the literature [33], under both user-user and item-item CF implementations. The DA_{next} algorithm has been found to consistently outperform all other algorithms in all tested datasets, reducing prediction errors, as reflected through the MAE and RMSE metrics, and also achieving the highest correct prediction percentages. In particular, the average MAE reduction compared to the plain CF algorithm is 5.1% under the user-user CF implementation and 6.0% under the item-item CF implementation; the respective gains regarding the RMSE metric are 5.3% and 7.1%. In comparison to the runner up algorithm, i.e. the $DA_{previous}$ algorithm [33], the improvements in the MAE are 1.5% under the user-user CF implementation scenario and 2.6% under the item-item implementation scenario, while the respective gains in the RMSE are quantified to 1.5% and 3.0%. Considering the correct prediction metric, the proposed algorithm outperforms the runner up algorithm ($DA_{previous}$) by 1.09% under the user-user CF implementation scenario and by 1.2% under the item-item implementation scenario. These benefits are realized at the expense of drops in coverage, which range from negligible to tolerable; coverage drops by 0.7% on average, being less than 2.2% in all cases. For datasets with higher density, in particular, practically no coverage loss occurs. We have also compared the improvements in accuracy achieved by the proposed algorithm against the corresponding improvements achieved by other algorithms exploiting temporal dynamics, and DA_{next} has been found to achieve the most substantial improvements, in all cases.

Our future work will focus on investigating alternative methods for computing the dynamic averages, as well as employing different dynamic average techniques for different users, depending on the timestamp distribution of their rating history. The adaptation of other similarity metrics, such as the Euclidian distance and the Manhattan distance [49], to exploit information regarding identified shifts in rating practices will be investigated.

The matrix factorization technique [32] and the fuzzy recommender systems approach [50] are also particularly interesting areas for further research on how shifts in rating practices can be accommodated in these approaches. For the matrix factorization technique, in particular, the approach of time-aware matrix factorization models [44, 45] will be studied.

Since the DA_{next} algorithm introduced in this paper targets shifts in user rating practices, which is a distinct aspect than those addressed in other approaches (e.g. interest shifts; decay of old-aged ratings; etc.), opportunities exist for combining the algorithm presented in our paper with algorithms from other categories, so that even higher accuracy improvements can be harvested; such combinations will be investigated in our future work.

Finally, exploring methods for decreasing the space overhead for the implementation of dynamic averages, considering the maintenance of dynamic averages at a coarser granularity than the individual rating, such as monthly or yearly, as well as decreasing the need for recomputing dynamic averages due to the arrival of new ratings (e.g. periodic recomputation or the consideration of ratings only in a specific temporal vicinity, when computing dynamic averages), will be explored.

References

1. Balabanovic, M., Shoham, Y.: Fab: content-based, collaborative recommendation. Commun. ACM **40**(3), 66–72 (1997)
2. Dror, G., Koenigstein, N., Koren, Y.: Yahoo! music recommendations: modeling music ratings with temporal dynamics and item taxonomy. In: Proceedings of the 5th ACM Conference on Recommender Systems (RecSys 2011), New York, NY, USA, pp. 165–172 (2011)
3. Schafer, J.B., Frankowski, D., Herlocker, J., Sen, S.: Collaborative filtering recommender systems. In: Brusilovsky, P., Kobsa, A., Nejdl, W. (eds.) The Adaptive Web. LNCS, vol. 4321, pp. 291–324. Springer, Heidelberg (2007). https://doi.org/10.1007/978-3-540-72079-9_9

4. Herlocker, J.L., Konstan, J.A., Terveen, L.G., Riedl, J.T.: Evaluating collaborative filtering recommender systems. ACM Trans. Inf. Syst. **22**(1), 5–53 (2004)
5. Li, L., Zheng, L., Yang, F., Li, T.: Modeling and broadening temporal user interest in personalized news recommendation. Expert Syst. Appl. **41**(7), 3168–3177 (2014)
6. Burke, R.: Hybrid recommender systems: Survey and experiments. User Model. User-Adap. Interact. **12**(4), 331–370 (2002)
7. McAuley, J.J., Pandey, R., Leskovec, J.: Inferring networks of substitutable and complementary products. In: Proceedings of the 21th ACM SIGKDD International Conference on Knowledge Discovery and Data Mining, Sydney, NSW, Australia, pp. 785–794 (2015)
8. McAuley, J., Targett, C., Shi, J., van den Hengel, A.: Image-based recommendations on styles and substitutes. In: Proceedings of the 38th International ACM SIGIR Conference on Research and Development in Information Retrieval, Santiago, Chile, pp. 43–52 (2015)
9. MovieLens datasets. http://grouplens.org/datasets/movielens/. Accessed 22 Sept 2017
10. Harper, F.M., Konstan, J.A.: The MovieLens datasets: history and context. ACM Trans. Interact. Intell. Syst. (TiiS) **5**(4), 19 (2015). Article No. 19
11. Zhou, Y., Wilkinson, D., Schreiber, R., Pan, R.: Large-scale parallel collaborative filtering for the netflix prize. In: Fleischer, R., Xu, J. (eds.) AAIM 2008. LNCS, vol. 5034, pp. 337–348. Springer, Heidelberg (2008). https://doi.org/10.1007/978-3-540-68880-8_32
12. Margaris, D., Vassilakis, C.: Pruning and aging for user histories in collaborative filtering. In: Proceedings of the 2016 IEEE Symposium Series on Computational Intelligence, Athens, Greece, pp. 1–8 (2016)
13. Liu, N.N., He, L., Zhao, M.: Social temporal collaborative ranking for context aware movie recommendation. ACM Trans. Intell. Syst. Technol. (TIST) **4**(1) (2013). Article No. 15
14. Bakshy, E., Rosenn, I., Marlow, C., Adamic, L.: The role of social networks in information diffusion. In: Proceedings of the 21st International Conference on World Wide Web, Lyon, France, pp. 519–528 (2012)
15. Koren, Y.: Factor in the neighbors: scalable and accurate collaborative filtering. ACM Trans. Knowl. Discov. Data (TKDD) **4**(1) (2010). Article No. 1
16. Vaz, P.C., Ribeiro, R., de Matos, D.M.: Understanding temporal dynamics of ratings in the book recommendation scenario. In: Proceedings of the 2013 International Conference on Information Systems and Design of Communication, ACM ISDOC 2013, New York, NY, USA, pp. 11–15 (2013)
17. Nishida, K., Yamauchi, K.: Detecting concept drift using statistical testing. In: Corruble, V., Takeda, M., Suzuki, E. (eds.) DS 2007. LNCS (LNAI), vol. 4755, pp. 264–269. Springer, Heidelberg (2007). https://doi.org/10.1007/978-3-540-75488-6_27
18. Liu, H., Hu, Z., Mian, A., Tian, H., Zhu, X.: A new user similarity model to improve the accuracy of collaborative filtering. Knowl.-Based Syst. **56**, 156–166 (2014)
19. Minku, L.L., White, A.P., Yao, X.: The impact of diversity on online ensemble learning in the presence of concept drift. IEEE Trans. Knowl. Data Eng. **22**(5), 730–742 (2010)
20. Elwell, R., Polikar, R.: Incremental learning of concept drift in nonstationary environments. IEEE Trans. Neural Netw. **22**(10), 1517–1531 (2011)
21. Ang, H.H., Gopalkrishnan, V., Zliobaite, I., Pechenizkiy, M., Hoi, S.C.H.: Predictive handling of asynchronous concept drifts in distributed environments. IEEE Trans. Knowl. Data Eng. **25**(10), 2343–2355 (2013)
22. Zliobaite, I., Bakker, J., Pechenizkiy, M.: Beating the baseline prediction in food sales: how intelligent an intelligent predictor is? Expert Syst. Appl. **39**(1), 806–815 (2012)

23. Vaz, P.C., Ribeiro, R., DeMatos, D.M.: Understanding temporal dynamics of ratings in the book recommendation scenario. In: Proceedings of the 2013 International Conference on Information Systems and Design of Communication (ISDOC 2013), Lisbon, Portugal, pp. 11–15 (2013)
24. Gama, J., Zliobaite, I., Bifet, A., Pechenizkiy, M., Bouchachia, A.: A survey on concept drift adaptation. ACM Comput. Surv. 1(1) (2013). Article No. 1
25. Margaris, D., Vassilakis, C., Georgiadis, P.: Recommendation information diffusion in social networks considering user influence and semantics. Soc. Netw. Anal. Mining 6(108), 1–22 (2016)
26. Ekstrand, M.D., Riedl, J.T., Konstan, J.A.: Collaborative filtering recommender systems. Found. Trends Hum.-Comput. Interact. 4(2), 81–173 (2011)
27. He, D., Wu, D.: Toward a robust data fusion for document retrieval. In: Proceedings of the 4th IEEE International Conference on Natural Language Processing and Knowledge Engineering (NLP-KE), Beijing, China, pp. 1–8 (2008)
28. Shardanand, U., Maes, P.: Social information filtering: algorithms for automating "word of mouth". In: Proceedings of the 1995 SIGCHI Conference on Human Factors in Computing Systems, Denver, Colorado, USA, pp. 210–217 (1995)
29. Margaris, D., Vassilakis, C., Georgiadis, P.: Knowledge-based leisure time recommendations in social networks. In: Alor-Hernández, G., Valencia-García, R. (eds.) Current Trends on Knowledge-Based Systems. Intelligent Systems Reference Library, vol. 120, pp. 23–48. Springer, Cham (2017). https://doi.org/10.1007/978-3-319-51905-0_2
30. Yu, K., Schwaighofer, A., Tresp, V., Xu, X., Kriegel, H.P.: Probabilistic memory-based collaborative filtering. IEEE Trans. Knowl. Data Eng. 16(1), 56–69 (2004)
31. Dias, R., Fonseca, M.J.: Improving music recommendation in session-based collaborative filtering by using temporal context. In: Proceedings of the 25th IEEE International Conference on Tools with Artificial Intelligence, Herndon, VA, pp. 783–788 (2013)
32. Koren, Y., Bell, R., Volinsky, C.: Matrix factorization techniques for recommender systems. IEEE Comput. 42(8), 42–49 (2009)
33. Margaris, D., Vassilakis, C.: Improving collaborative filtering's rating prediction quality by considering shifts in rating practices. In: Proceedings of the 19th IEEE International Conference on Business Informatics, Thessaloniki, Greece, vol. 01, pp. 158–166 (2017)
34. Bao, J., Zheng, Y., Mokbel, M.: Location-based and preference-aware recommendation using sparse geo-social networking data. In: Proceedings of the 20th International Conferences on Advances in Geographic Information Systems (SIGSPATIAL 2012), Redondo Beach, California, pp. 199–208 (2012)
35. Zheng, Y., Xie, X.: Learning travel recommendations from user-generated GPS traces. ACM Trans. Intell. Syst. Technol. (TIST) 2(1), 29 (2011). Article No. 2
36. Gong, S.: A collaborative filtering recommendation algorithm based on user clustering and item clustering. J. Softw. 5(7), 745–752 (2010)
37. Das, A., Datar, M., Garg, A., Rajaram, S.: Google news personalization: scalable online collaborative filtering. In: Proceedings of the 16th International Conference on World Wide Web, Banff, Alberta, Canada, pp. 271–280 (2007)
38. Margaris, D., Vassilakis, C.: Enhancing user rating database consistency through pruning. Trans. Large-Scale Data Knowl.-Centered Syst. XXXIV, 33–64 (2017)
39. Ramezani, M., Moradi, P., Akhlaghian, F.: A pattern mining approach to enhance the accuracy of collaborative filtering in sparse data domains. Phys. A: Stat. Mech. Appl. 408, 72–84 (2014)
40. Najafabadi, M.K., Mahrin, M.N., Chuprat, S., Sarkan, H.M.: Improving the accuracy of collaborative filtering recommendations using clustering and association rules mining on implicit data. Comput. Hum. Behav. 67, 113–128 (2017)

41. Li, C., Shan, M., Jheng, S., Chou, K.: Exploiting concept drift to predict popularity of social multimedia in microblogs. Inf. Sci. **339**, 310–331 (2016)
42. Lu, Z., Pan, S.J., Li, Y., Jiang, J., Yang, Q.: Collaborative evolution for user profiling in recommender systems. In: Proceedings of the 25th International Joint Conference on Artificial Intelligence (IJCAI 2016), pp. 3804–3810 (2016)
43. Kangasrääsiö, A., Chen, Y., Głowacka, D., Kaski, S.: Interactive modeling of concept drift and errors in relevance feedback. In: Proceedings of the 2016 Conference on User Modeling Adaptation and Personalization (ACM UMAP 2016), New York, NY, USA, pp. 185–193 (2016)
44. Gantner, Z., Rendle, S., Schmidt-Thieme, L.: Factorization models for context-/time-aware movie recommendations. In: Proceedings of the Workshop on Context-Aware Movie Recommendation (ACM CAMRa 2010), New York, NY, USA, pp. 14–19 (2010)
45. Zhang, J.D., Chow, C.Y.: TICRec: a probabilistic framework to utilize temporal influence correlations for time-aware location recommendations. IEEE Trans. Serv. Comput. **9**(4), 633–646 (2016)
46. Sarwar, B., Karypis, G., Konstan, J., Riedl, J.: Item-based collaborative filtering recommendation algorithms. In: Proceedings of the 10th International Conference on World Wide Web (WWW10), Hong Kong, pp. 285–295 (2001)
47. Winoto, P., Tang, T.Y.: The role of user mood in movie recommendations. Expert Syst. Appl. **37**, 6086–6092 (2010)
48. DELL: PowerEdge R930 Rack Server specs. http://www.dell.com/us/business/p/poweredge-r930/pd?ref=PD_OC. Accessed 22 Feb 2018
49. Cha, S.-H.: Comprehensive survey on distance/similarity measures between probability density functions. Int. J. Math. Models Methods Appl. Sci. **1**(4), 300–307 (2007)
50. Son, L.H.: HU-FCF: a hybrid user-based fuzzy collaborative filtering method in recommender systems. Expert Syst. Appl. **41**, 6861–6870 (2014)
51. Guo, G., Zhang, J., Thalmann, D., Yorke-Smith, N.: ETAF: an extended trust antecedents framework for trust prediction. In: Proceedings of the 2014 International Conference on Advances in Social Networks Analysis and Mining ASONAM 2014, Beijing, China, pp. 540–547 (2014)
52. Lo, Y.-Y., Liao, W., Chang, C.-S., Lee, Y.-C.: Temporal matrix factorization for tracking concept drift in individual user preferences. IEEE Trans. Comput. Soc. Syst. **5**(1), 156–168 (2018)
53. Cheng, W., Yin, G., Dong, Y., Dong, H., Zhang, W.: Collaborative filtering recommendation on users' interest sequences. PLoS One **11**(5), e0155739 (2016)

Author Index

Bennani, Nadia   41
Bollwein, Ferdinand   1
Brunie, Lionel   41

Christodoulou, Klitos   81

Fernandes, Alvaro A. A.   81

Gerl, Armin   41

Kosch, Harald   41

Margaris, Dionisis   151

Norta, Alex   113

Paton, Norman W.   81

Qayumi, Karima   113

Serrano, Fernando Rene Sanchez   81

Vassilakis, Costas   151

Wiese, Lena   1

Printed in the United States
By Bookmasters

Printed in the United States
By Bookmasters